How Deeply Human Is Language?

How Deeply Human Is Language?

Chomsky, the Brain, and the AI Fantasy

Yosef Grodzinsky

The MIT Press
Cambridge, Massachusetts
London, England

The MIT Press
Massachusetts Institute of Technology
77 Massachusetts Avenue
Cambridge, MA 02139
mitpress.mit.edu

The MIT Press would like to thank the anonymous peer reviewers who provided comments on drafts of this book. The generous work of academic experts is essential for establishing the authority and quality of our publications. We acknowledge with gratitude the contributions of these otherwise uncredited readers.

This book was set in Stone Serif and Stone Sans by Westchester Publishing Services. Printed and bound in the United States of America.

Library of Congress Cataloging-in-Publication Data is available.

ISBN: 978-0-262-05200-9

10 9 8 7 6 5 4 3 2 1

EU Authorised Representative: Easy Access System Europe, Mustamäe tee 50, 10621 Tallinn, Estonia | Email: gpsr.requests@easproject.com

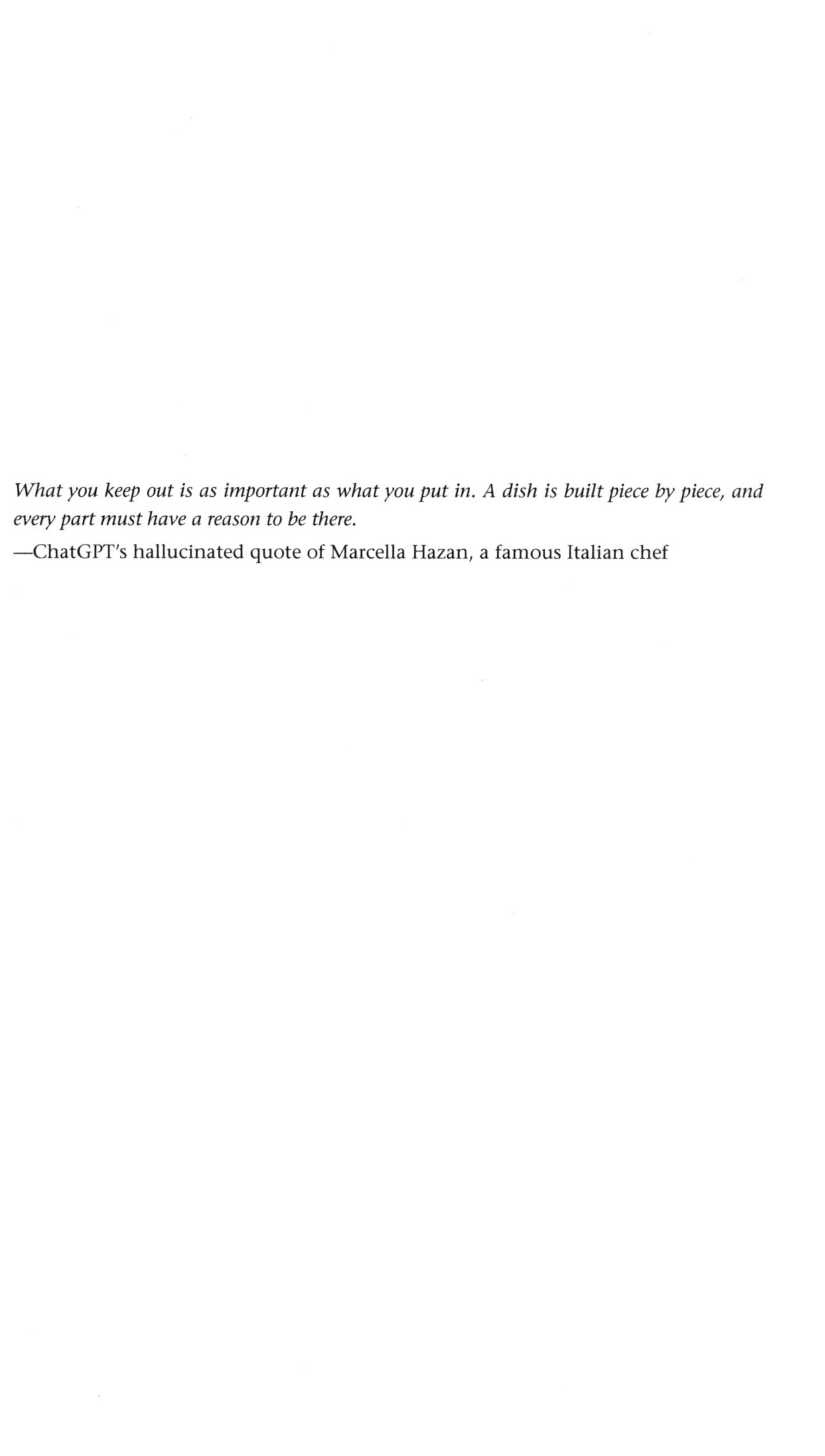

What you keep out is as important as what you put in. A dish is built piece by piece, and every part must have a reason to be there.

—ChatGPT's hallucinated quote of Marcella Hazan, a famous Italian chef

Contents

Preface and Acknowledgments

Two huge events, a decade apart from one another, have pushed the technological world in new and unprecedented directions: the 2012 presentation of AlexNet, a revolutionary image recognition system, and the 2022 release of ChatGPT. These innovations, as well as the groups of AI experts who developed them, gave the world a feeling that a new era is here, one that eclipses all past industrial revolutions combined.

At first, images and other visual materials were at the heart of the public discussion of AI; at present, language seems to be centerstage. And here, linguists and linguistics—who have largely remained outside of the engineering enterprise—may have something to contribute. And yet, with few exceptions (e.g., Emily Bender, Roni Katzir, and Gary Marcus), linguists have hardly engaged in any real discussion or debate. A long-time veteran of linguistics and neurolinguistics, I felt fit to jump in, hoping to offer something new. A plan followed, and a generous sabbatical from my home institution enabled me to retreat, first to a huge metropolitan center in Argentina (Buenos Aires), and later to a small town in Germany (Monheim am Rhein). Before long, a book was brewing. I ended up telling a story that begins on October 8, 2024, ends a year earlier to the day, and tours nearly two centuries in the middle.

Writing a book is a marathon—easy at the beginning and yet scary, as your heart is full of self-doubt about whether you'll make it to the finish line; it gets routinized and rhythmic in the middle; nearer the end, you know it's yours, but it becomes an unbearably painful and difficult experience. You stick to it, but unlike marathons, no certificates or medals are handed out at the finish line.

I used to be a reasonable sprinter, but I always was a bad long-distance runner. The pain was just too intense for me to bear. But in writing this book, the pain I endured was converted into joy, thanks to friends, colleagues, and relations on three continents, who came to my rescue as the going got tough. Many were there for me at the crucial junctures: Katrin Amunts, Sebastian Bludau, Amy Brand, Ronit Calderon-Margalit, Anat Caspi, Yinon Cohen, Luka Crnič, Lotem Elber, Danny Fox, Julieta Fumagalli, Virginia Jaichenco, Leo Joskowicz, Gili Laor, Thomas Lippert, Yonatan Loewenstein, Dany Margalit, Shai Moise, Yehonatan (Hilik) Nadav, Eli Nelken, Peter Pieperhoff, Marion Rapp, Beth Rosenberg, Michael Sabel, Anat Saragusti, and Niza Yanay. I deeply appreciate their help and friendship, and I thank them from the bottom of my heart.

I am also most grateful to my computational neuroscientist colleagues at the Edmond and Lily Safra Center for Brain Sciences (ELSC) for strongly disagreeing with me and for asking hard questions, both in person and in correspondence, especially David Beniaguev, Jonathan Kadmon, Yonatan Loewenstein (once again), Mickey London, Aviv Mezer, Idan Segev, and Haim Sompolinsky. Entering this lions' den was most beneficial.

Many thanks go to five helpful MIT Press reviewers, to the Press's editorial staff, to I-An Tan who made the index, and to ELSC's internal grant support, combined with the generous funding received from the European Union's Horizon Europe program under Specific Grant Agreement 101147319 (the EBRAINS 2.0 project) by the Institute for Neurosciences and Medicine (Structural and Functional Organization of the Brain division) of Forschungszentrum Jülich.

Additional thanks go to the audiences in various courses, colloquia, and summer schools on three continents, whose questions and comments helped me sharpen my thinking.

And of course, I would never have written this book without the support of those closest to my heart. Amalya Grodzinsky-Liebermann was there at every juncture and generously contributed not only her love, but also graphic art and wise advice; and Irit Frank was there lovingly, both for me and with me, every single second. I know that I've been very lucky to have them, and I am most grateful to them for their love and support.

Prologue: Is Bigger Better?

ANNIE: *Anything you can do, I can do better.*
I can do anything better than you.
FRANK: *No you can't.*
ANNIE: *Yes, I can.*
FRANK: *No, you can't.*
ANNIE: *Yes, I can.*
FRANK: *No, you can't.*
ANNIE: *Yes, I can, yes, I can!*
—"Anything You Can Do," Irving Berlin, *Annie Get Your Gun*, 1950

The biggest lesson that can be read from 70 years of AI research is that general methods that leverage computation are ultimately the most effective, and by a large margin.
—Rich Sutton, "The Bitter Lesson," 2019

A Nobel Interview in the Wee Hours, October 8, 2024

On October 8, 2024, at 11:45 a.m. Central European Time, the Physics Nobel Prize Committee announced that this year's prize was to be awarded to Geoffrey Hinton—a professor of computer science at the University of Toronto and laureate of many prestigious prizes—alongside Princeton's John Hopfield. The Committee noted that Hinton had "used the Hopfield network as the foundation for a new network that uses a different method: the *Boltzmann machine*. This can learn to recognise characteristic elements in a given type of data. Hinton used tools from statistical physics, the science of systems built from many similar components. The machine is trained by feeding it examples that are very likely to arise when the machine is run. The Boltzmann machine can be used to classify images or create new examples of the type of pattern on which it was trained. Hinton has built upon this work, helping initiate the current explosive development of machine learning."[1]

Shortly after the announcement, Hinton, who was housed in a "cheap hotel in California," as he described it, received another call, this time from Adam Smith of the Nobel Prize website. It was about 3 a.m. Pacific Time, and Smith was calling for the customary "First Reactions" brief interview.[2] The interview went quite well. Here is its final minute:

> SMITH: *So, for instance with the large language models, the thing that I suppose contributes to your fear is you feel that these models are much closer to understanding than a lot of people say. When it comes to the impact of the Nobel Prize in this area, do you think it will make a difference?*
>
> HINTON: *Yes, I think it will make a difference. Hopefully, it'll make me more credible when I say these things really do understand what they're saying.*
>
> SMITH: *Do you worry that people don't take you seriously?*
>
> HINTON: *So, there is a whole school of linguistics that comes from Chomsky that thinks that it's complete nonsense to say these things understand, that they don't process language at all in the same way as we do. I think that school is wrong. I think it's clear now that neural nets are much better at processing language than anything ever produced by the Chomsky School of Linguistics. But there's still a lot of debate about that, particularly among linguists.*

Many found this text somewhat shocking. Is Hinton's field of view from the very top of the world marred by yet another peak that must be conquered? Is the superiority of large language models (LLMs) so fragile that it must be stipulated at this sweet moment? What is the nature of the debate Hinton is alluding to? Is it not overly ambitious to expect, indeed to claim, that a physical model that classifies images well can also understand language?

This book is an extended meditation on these questions. A scientific account of the linguistic abilities we humans seem to possess has long been sought after. The engineering effort to build talking machines is familiar to every reader of this book, whether through its best current incarnation, ChatGPT, or through its tributaries and offspring. Many readers may also be aware of Noam Chomsky's words, before he went silent, to the effect that like bulldozers, deep learning devices such as Google Translate are useful, but they tell you nothing about what human language is. Whence this debate, and what exactly is it about? In this book, I address these questions in detail, describe the state of the art on both sides, explain the ideas behind each approach, highlight their accomplishments and problems, and conclude by making some proposals for a better future.

Telling a Good Story

First, we must listen to the story as told by both sides—we humans need stories that motivate us to persist and persevere. And here, there are two stories: one, *linguistic theory*, is told by scientists; the other, *LLMs*, is told by engineers. Linguistic theory and linguistics as a field have had major accomplishments over the past 70 years, but these have not lent themselves easily to the development of practical tools, and they have often been scorned as being useless; LLMs' language application developers and AI as a field build great devices and tools, but they have found themselves accused of putting the practical before the theoretical, of having biases (gender, race, and perhaps other), of condoning exploitative conduct, and of developing monstrous machines that are a threat to the future of humanity.[3]

Each story claims exclusive rights, as we shall see. Which, if either, should we believe? This book explores the current state of scientific and technological inquiries into human language. The public discussion of this hot topic has touched on many important points,[4] but naturally, certain critical questions have remained veiled. Using nontechnical language, I first offer a tutorial in which I present basic facts and explain basic issues underlying each position, and I follow by putting these critical questions on the table, in an attempt to launch an open discussion. As a cognitive neuroscientist who is also trained in formal linguistics, I focus on highly specific issues and bring up—in nontechnical language—a host of novel arguments that rely on clearcut empirical considerations and experimental results. I also try to come up with some concrete answers and some modest proposals for a better future. Let me offer a brief sketch of the two stories.

The Linguistic Story

Up until the end of World War II, the academic study of language was mostly engaged with exciting stories of the past or the exotic: European philologists studied Old Church Slavonic palimpsests, the declension of nouns in

Attic Greek, and the etymology of *strozzapreti*, that famed Northern Italian pasta, while their North American counterparts investigated the dialects of Ojibwe or the concept of time in Hopi. These collections of observations and facts, drawn from varied human activities, fascinated scholars of old texts, as well as cultural historians and anthropologists.

After the war, the language sciences underwent dramatic changes. Mathematical tools gave way to grammatical formalisms, which paved the way for Noam Chomsky's dazzling appearance on the intellectual scene. Chomsky had a new story to tell, one that revived ancient questions: human language, he said, is not just a bag of words but a formal system of grammar rules of various types. We as speakers know these rules, which leads to the conclusion that they are part of our mental existence. Linguistics is therefore the study of a critical ingredient of human mental structure. In making this argument, Chomsky placed linguistics at the heart of the philosophical and psychological investigation of the human mind, which naturally connected to questions of cognitive development in childhood, the dissolution of cognitive faculties subsequent to brain disease, and a host of related psycho- and neurobiological issues.

Capturing the imagination of many, Chomsky proposed a new research program, and with the help of Morris Halle, Jerome Wiesner, and several others, a graduate program in the new linguistics was founded at MIT in 1961, with a focus on formal linguistic theory in all its manifestations, from phonetics to morphophonology and syntax and, later, to semantics and pragmatics. Under Chomsky and Halle's tutelage, the theory developed in important and interesting ways, both substantive and methodological, discovering a huge body of new facts and puzzles about language and proposing a highly structured theory with myriad abstract generalizations that explain these puzzles. Generations of linguists have brought the theory to where it is today—light years ahead of where it started. These advances have enhanced our understanding of language, our humanity, and, to an extent, the linguistic brain, shedding light on certain intricate patterns of experimental results. And yet, hardly any of the highly sophisticated formal theories that theoretical linguists have developed are successfully implemented in language technologies. The industry thus turned away from the story told by linguists and connected to another exciting story.

The Language Engineering Story

Since at least the 1940s, neurophysiologists, mathematicians, and engineers have tried to realize the dream of modeling the central nervous system with neuron-like networks that can learn; in parallel, first attempts at creating mechanical translation machines were made. Three-quarters of a century later, and following many false starts, the two were combined in LLMs, whose crown jewel, ChatGPT, shocked the world. This amazing machine, which seems to have been created mostly by smart Millennials who are good at math, has been immediately put to use by Gen Z kids, and actually by everyone else too. As every reader of this book knows, ChatGPT makes you feel that you're conversing with a smart and knowledgeable friend, a human just like you, though a faster and better informed one. ChatGPT in its various incarnations is thus the present chapter of several long-standing traditions, which were rolled into a new, attractive story—an artificial simulation of the human brain that behaves like a human does and that, so the story goes, accounts for findings from "wet" brain experiments. This Herculean R&D effort by industrial agents has thus raised hopes that an understanding of one of humanity's greatest mysteries—our ability to communicate through language—is within reach.

The Grim Way Things Are

These two approaches to language and communication have turned out to be vastly different. The stories they tell rely on different mathematical machineries, they find different language tasks important, their products are quite distinct, and if they make new discoveries, these are of very different kinds. The juxtaposition between them raises several questions: How adequate is the account of language given by each? Can linguistic theory help the development of LLMs (e.g., by trimming massive power-hungry computations, thereby saving them time and electricity—a major issue that has led some to suspect LLM development is about to hit a wall)? Can LLMs help linguistic theory (e.g., by delivering better implementations of language acquisition and processing algorithms)? Haven't we fallen victim to an *implementation bias*, which prefers running computer programs to theories that may have advantageous properties other than "runnability"?

The name of this book suggests a preference (though not a bias) on my part: as a scientist, my main inclination is to look for the scientific merits of a theoretical account. Technological victories are important and extremely hard to accomplish, but they do not necessarily explain our humanity. Throughout the book, I will try to draw a clear line between technology and science (although admittedly, such a line may not always exist). The questions to be addressed here, then, are hard and subtle, and I will try to provide clear and detailed answers. Toward the end, I will touch on further puzzles that have more of a sociological flavor: Do the two approaches currently benefit from each other's toolbox? Is there cross talk between the two communities? Can a combined, interwoven account work better? These questions are critical to both science and industry. Beyond answers, I will try to make some modest proposals. But I can already provide an unhappy hint: something seems to have gone wrong, as sadly, there is hardly any dialogue between the language engineering and theoretical linguistics communities. If anything, there is mutual scorn and bile, coming out of leaders such as Geoffrey Hinton, who as noted has won the 2024 Nobel Prize in Physics and who, in addition to many others, recently stated that "Chomsky managed to convince his followers that language wasn't learned" (a plainly false attribution) and has repeatedly called Chomsky "crazy."[5] Hinton and others have taken stabs at Chomsky but not so much at the theory he and many subsequent generations of theoretical linguists have developed. On the "other side," the same Noam Chomsky—my revered mentor—also had his share of sweeping dismissals of AI.[6]

These dismissive and divisive comments, coming from leaders on both sides of the aisle, are most regrettable and not helpful. No wonder, then, that hardly any joint work is ever initiated, to everyone's detriment. To me, a dialogue is prerequisite to discovery and progress, hence a must. "The key word here is *encounter*," writes food historian Massimo Montanari in his recent *Short History of Spaghetti with Tomato Sauce*. "The more numerous and more interesting the encounters, the richer the result, the stronger and more robust the plant." I could not agree more.

The Critique in Brief

The quote from ChatGPT at the beginning of this book is coherent and grammatically well-formed. Indeed, LLMs are touted as "revolutionary" devices, whose development is an unprecedented "paradigm shift." But this quote also turned out to be fabricated, "hallucinated" by the machine. While I will not offer a general critique of AI (such as the one in Gary Marcus's recent *Taming Silicon Valley*),[7] I will reflect on such oddities and their relevance to the evaluation of LLMs as models for human language behavior. The hard look I will take at these models from a (neuro)linguist's perspective will expose deeply deficient practices: the preference of prediction to structured and constrained explanation; the belief in brute force computation (eloquently expressed by Sutton's quote above), even if it entails the construction of hyperpowerful machines whose nontransparent internal structure is difficult to analyze, and consequently might be doing much unnecessary computation that consumes unimaginable amounts of precious energy; the unreasoned exclusion of apparently relevant datasets; the existence of concealed biases; deficient evaluation abilities through inadequate benchmarks; absence of theoretical constraints; and a frequent failure to consider alternatives.

I will not shy away from criticizing linguists' practices, which also tend at times to be problematic, albeit perhaps to a lesser degree. In some though not all theoretical linguistic communities, one notices a lack of interest in highly relevant empirical data, typically in the results of psycholinguistic and neurolinguistic experiments; moreover, as a field, theoretical linguistics has maintained weak standards regarding quantitative methods for data analysis. Finally, linguists have been slow in gearing themselves toward implementations and useful practical applications. All of these have also been quite detrimental to the development of the field.

The discussion to come, then, will cover both sides. It will be bidirectional, though not symmetric. I will conclude by sketching a cooperative horizon, which promises to advance both science and engineering.

An Alternative and Its Merits

A critique of an approach is most valuable and compelling when accompanied by an alternative proposal, or at least a sketch thereof. In the final chapters of the book, I will present an approach that is both empirically grounded and based on methodological considerations. After demonstrating what has gone wrong, I will make a general proposal: to replace the current gigantic LLMs with small, highly constrained and modular networks.

Then, I will show how this proposal for the modeling of language is very much in the spirit of current modeling of mammalian vision and in keeping with what we know about the mosaic of language areas in the human brain.

What's in the Box?

Let me now go briefly over the book's content, by parts and chapters, and sketch the issues I discuss. Following this prologue, the book has four two chapter parts and an epilogue. In Part I, "Where We Are: The Language Sciences," I tell the story of generative grammar, its origins, development, and current state. Chapter 1, "Grammars, Old and New" I describe the beginning of grammar, its development over the ages up to the 1950s, and the intellectual climate that gave rise to Noam Chomsky's earliest proposals of transformational generative grammar (TGG), which I sketch and rationalize. Chapter 2, "Hints from Present-Day Linguistic Theory," gives a quick nontechnical tour of TGG's current state of the art, at a level that covers multiple (at times competing) linguistic frameworks. In both chapters, I tell the story of the theory and its main insights via presentations of specific issues in syntax and semantics, which concerns utterances that should be close to any speaker's (and reader's) heart and mind. I try to emphasize the problems, the logical structure of their proposed solutions, and the implications these may have for a general explanation of the human linguistic ability. Importantly, I present the theory while minimizing technical issues. This approach not only has the advantage of being inclusive of multiple linguistic perspectives and styles, but also, it assumes no prior knowledge of linguistics. Part II, "Where We Are: Language Technologies," gives a similar treatment to LLMs. Chapter 3, "From Early Perceptrons to Deep Learning," describes the history of neural networks from its anatomic-physiological beginnings in the early twentieth century to the work of the first modelers of neurons in the 1940s through the late 1950s' learning rule and all the way up to deep learning and the dramatic appearance of AlexNet in 2012. Then chapter 4, "Dreams About Talking Machines: Current LLMs," explains (again, nontechnically and requiring no prior knowledge) what networks are and what they can do. Part III, "Getting Under the Hood," is a juxtaposition, beginning with chapter 5, "A Duel: Science Versus Technology," in which the conceptual underpinnings of the theory of TGG is compared to LLMs. In it, issues such as the Innateness Hypothesis, constrained versus unconstrained rule systems, and other conceptual issues are discussed, with concrete case studies. Chapter 6, "Poking Linguistic Holes in LLMs (ChatGPT Included)," takes a hard look at ChatGPT and its kin. Unlike many, I do not engage in a cat and mouse chase of the errors, hallucinations, confabulations, and biases of this impressive machine. Rather, I try to locate fundamental problems, and reflect on why they are there (hint: lack of constraints or linguistic structure in the underlying networks). Part IV, "The Brain's Language Code," is dedicated to linguistic theory and LLMs in the context of the human brain. This part is purely about science—about the best account of the way the human brain talks and understands. Its first chapter 7, "A Mosaic of Neurolinguistic Modules," reviews what we know about the functional structure of brain parts that are entrusted with linguistic analysis. Providing a linguist's perspective on brain–language relations, this chapter presents weighty hints—from experiments with brain-damaged individuals, from functional neuroimaging in healthy participants, and from direct measurements of brain activity—in favor of the view that the modular nature of linguistic theory is reflected directly in brain structure. That is, specific brain parts carry out specific linguistic computations, in a manner dictated by TGG. The counterpoint chapter 8, "Modeling the Linguistic Brain with LLMs," presents the LLM perspective, by which a certain class of neural network models is said to account for all language performance, as measured directly from the human brain. Here, too, the general view is bolstered by experimental results from brain measures that are similar to the previous ones, although the conclusion is quite different. The two perspectives may appear incommensurate, and each has problems, which will be discussed.

In the epilogue, "Shall We Work Together?," I consider ways for us to move forward, beyond division, sectarianism, and powerplay. Instead of two competing factions playing against each other, we need to team up in a joint effort to understand the nature of human linguistic ability. And while I can already disclose that my account will not be entirely symmetric and will be more inclined toward the linguistic side, it is clear to me that neural modeling and implementation are critical ingredients of a theory that strives to give us deeper understanding of brain mechanisms for language.

How to Read This Book

I have aimed to reach a broad spectrum of readers: people interested in philosophy, science, and technology, curious readers who have heard about modern linguistics and about Noam Chomsky but have never understood what the fuss was about; engineering students who have taken or plan to take a course in natural language processing; linguists who fear that their discipline may soon be erased off the academic map; neurologists and neurosurgeons who have seen the wonders of the talking brain, as well as speech and language therapists who try to fix it when it fails; engineers and developers who may have followed some of the debates on X (formerly Twitter), YouTube, and other media, and want to build better language bots; and, of course, my academic colleagues from all relevant disciplines. There's something here for everyone, I hope.

The book touches on technical issues and tries to do so in a nontechnical way. Nevertheless, it tries to present complicated and subtle theoretical and conceptual issues from a range of fields. Occasionally matters become a bit hairy in a way that requires slow and reflective reading, which may be anathema to some. The somewhat more difficult parts are presented in boxes. Everything in the introductory part is used later. My advice: read every part as far as you can, but do keep in mind that skipping to the next chapter (and hopefully coming back at some later stage) is an option. It is not a detective book. You can go back and forth and, sometimes, just move on and never come back. As long as you get the gist of things, I'll be pleased.

Happy reading!

Part I Where We Are: The Language Sciences

1 Grammars, Old and New

When I pronounce the word Future,
the first syllable already belongs to the past.
—Wisława Szymborska

The capacity to be puzzled about what looks obvious is a good property to cultivate.
—Noam Chomsky, comment during panel discussion at MIT, 2011

The Beginning of Grammars

In the beginning of grammars, there was Pāṇini. The first linguistic hero, he wrote a grammar of Sanskrit that has been celebrated as the earliest ever. Pāṇini may have belonged to a first generation of Brahmans who could write,[1] and observations on the difference between spoken and written Sanskrit led him to build a system of grammar to describe these differences. The resulting *Aṣṭādhyāyī* ("Eight Chapters," written ~400–300 BCE) proposed a complete, concise, and consistent analysis of grammatical structure, through thousands of rules and a large inventory of phonological elements and verbal roots, as well as long lists of "exceptions." Coming from India, it is the earliest known attempt to introduce precision, systematicity, consistency, and conciseness into the study of human language.[2]

Everyone thinks about language, and everyone has an opinion about its nature. But what is grammar, and what, if anything, does it tell us about ourselves? What exactly did Pāṇini, and a multitude of grammarians and linguists who followed, write? Pāṇini was struck by the richness and regularity of the sounds and sound combinations in Sanskrit, and he constructed a system of rules to captured them. From then on, grammarians have held that a grammar reflects relations between elements of language (speech sounds, syllables, words, phrases, sentences) as determined by its native speakers (who master the grammatical intricacies of their language without having ever been explicitly taught its grammar).

But what does a rule look like? Let's consider a simple English example from the formation of plural (PL.) nouns: dog+PL. → *dogz*; *cat*+PL. → *cats*; *dress*+PL. → *dressez* (or *dresseez* if you're British). These rules are simple indications of how to combine a word (*dog*) with an element of meaning (PL.) and map them ("→") onto another word (*dogz*). But writing a plural rule for each word would not get us very far, as Pāṇini understood. Instead, we should try to understand the generalization behind these specific mappings, and construct a rule that would indicate what is permitted and what isn't. The English plural rule harbors a principle that most readers know: assume that the plural *-s* is added to a word, but that it can be realized in different ways, depending on the features of the consonant that precedes it. If its pronunciation requires the immediate vibration of the vocal cords (known as "voicing," as in *b, d, g, v*), then the immediately subsequent plural *-s* inherits this property (hence *dogz*); if the word ends with an "unvoiced" consonant (like *p, t, k, f*, in which vocal cord vibration is delayed), then the plural *-s* remains as is (*cats*); and if the final consonant of the word is *-s* or *-z*, then pluralization would be *-ez* (*dressez*; alternately *-eez*, hence *dresseez*). Once formulated thus, we can inquire whether this rule fits into a general mold of English pronunciation: is it in fact related to plurals at all, or does it instead characterize a more general proximity effect, involving two consonants that belong to distinct pieces of meaning, regardless of pluralization? It is clear that the underlying principle has little to do with plurals; rather, the pronunciation of plurals follows a more general rule, one that ties the proximity of an added piece of meaning to the consonant that precedes it, not only in nouns. This is why we say *he hop-s and pop-s* but *she rob-z and cuss-ez*, which are pronounced in keeping with the same rule, despite their being verbs. The result is more general than the previous one, but it still doesn't cover everything, for what about *he pay-z* or *she fall-z*, in which a final *-s* is not preceded by a voiced consonant? Are these exceptions, or did we misstate the rules? And maybe we got the rule backwards and its default is *-z*, not *-s*? That is, perhaps *dogz*→*dog*+PL. is a better description of this particular instantiation of the pluralization rule? These are serious questions that linguists grapple with, attempting to come up with reasons to prefer one approach to others, thinking that the best (i.e., most general and precise) formulation is the one that native English speakers tacitly hold in their heads. And we haven't even gotten to question the rule's validity across languages, or to wonder about general laws of pronunciation and their cognitive and neurophysiological bases.

There is much to talk about, but let's turn to what grammatical rules imply. In a way, they define the limits of language. A native English speaker can say *he hop-s and pop-s* and *she rob-z and cuss-ez*, but not **he hop-z and pop-ez* or **she rob-s and cuss-s*. The same holds for the possessive -*'s*: we say *the dog'z and the cat's tail* but not **dog's*, *cat'z. The latter are plainly not within the boundaries of English, as judged by its native speakers (hence we mark them with the "*" symbol). The final -*s* rule, then (and grammar in general), codifies regularities in language and draws the line between possible linguistic objects and impossible ones (**dog-s*, **hop-z*, **dog's*). And as we have seen, what began as a rule for plural nouns can be stated at different levels of abstraction.

Let me pause here to note that we are not looking at picayune little things of no consequence. The English plural/third-person/possessive -*s* is small, but like particles in physics, small may be quite important. Indeed, if the -*s* rule is marginal, how come its effects are so far-reaching? Every native speaker of English abides by it: nobody ever utters any of the starred words above. This important point—that grammar marks the boundaries of language and excludes incorrectly structured words we would never utter—is not limited to words; bigger linguistic units also abide by grammatical rules. Aristotle, who lived around Pāṇini's time, famously noted this in his *On Interpretation*, in which he presented a system of logic. While reasoning, not grammar, was his focus, he analyzed sentences into what he thought were their component parts. Indeed, *On Interpretation* is sprinkled with allusions to familiar sentence level concepts, such as *subject* (to Aristotle, the main player in a sentence, the qualities of which the rest of the sentence characterizes), *predicate* (the activity or state that characterizes the subject; usually a verb), *adjective* (a descriptor of a subject or object's qualities, such as *happy*, *tall*, *smart*), and many other terms that indicate the author's interest in grammatical structure (arguably, such an interest is already evident in Plato's *Cratylus*). Grammar and language structure, then, were prominent concepts utilized by thinkers in the ancient world. And we have clear evidence for continued interest in grammar during the Middle Ages, in the form of French, Arabic, Hebrew, and Latin grammars,[3] and the list goes on and on.

To us, the real story begins in the twentieth century, with the birth of modern linguistics, especially in the United States. There, Leonard Bloomfield developed a method to analyze sentences into their *immediate constituents*, a system in which a sentence like *The big dog chased the small cat* is divided into words, then phrases—pairs of "immediately" adjacent words that "go best together"—then larger phrases made out of pairings of phrases with their neighbors, and so on up to the top constituent: the sentence. This is shown in figure 1.1 (where each pairing is marked by two lines that meet).[4]

Constituency implies a hierarchy in relations among elements. The existence of intermediate branches, or nodes, indicates which parts are more closely related to which—in this sentence, the words *small* and *cat* form a constituent (*[small cat]*), and the word *the* forms a constituent with it (*[[the][small cat]]*). Thus, *the* is higher than each of the words *small* and *cat* (in a sense that must be made precise), as it combines with the constituent they form together. This comports with our intuition: what is modified by *the* is neither *small* nor *cat* but their combination. Our linguistic intuition also seems to suggest that *[chased the]* is an arbitrary grouping of words; and indeed, the diagram in the figure (as well as the equivalent bracketing in the text) indicates that this pair of words is not a constituent, because there is no node that contains all these words, and only them. This conclusion also fits with the results of certain tests: for example, you can assert, emphatically, that *it is* ***the small cat*** *that the big dog chased,* but English would not tolerate **it is* ***the cat*** *that the big dog chased* ***small,*** which is a plain monster. That is because *the small cat* is a constituent and must appear in one piece.

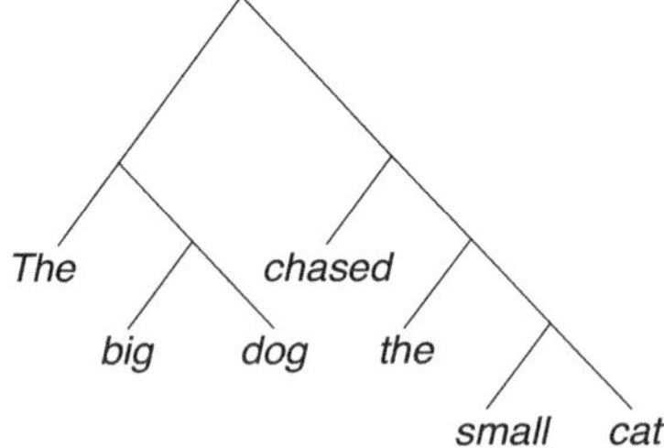

Figure 1.1
A sentence parsed into its immediate constituents.

This system has some transparency, enabling its proponents to justify every one of their moves. For instance, how exactly is the diagram in figure 1 drawn? Why is the division binary, not otherwise? What is the significance of a hierarchy? Linguists have been deeply concerned with these questions. While Bloomfield's insight was that this hierarchy is important for our understanding of how language works, Zellig Harris constructed a formal system of rules that built sentences out of immediate constituents and thus captured many linguistic regularities. Importantly, Harris's formal system captured phenomena but offered no constraints (a critical issue, on which I will elaborate).[5] The linguistic scene was now ready, as it were, for the main language event of the mid-twentieth century: the appearance of generative grammar, proposed by Noam Chomsky.

The Zeitgeist in the Language Sciences, Early 1950s

We are all captives of the intellectual landscape into which we were born and raised. During the early 1950s, the community of those engaged in the study of language and related behaviors was divided into four prongs: the *structural*, the *statistical*, the *behavioral*, and the *cultural*.

Structuralists, such as Bloomfield and Harris, tried to build a hierarchical grammatical system that would distinguish well-formed words and utterances from monsters and would characterize them precisely, mechanically, and reliably. At the word and sentence level, Harris introduced mathematical methods into the analysis of hierarchical relations in sentence level grammars, but no constraints on grammatical systems were formulated—the linguist could fit rules to a language and build its inventory of "trees" with no limits. As a result, there was no understanding of what a possible syntactic rule of a human language is or of why languages differ from one another.

Parallel developments took place in phonetics. Speech scientists, who investigated the sound shape of language and speech, introduced sophisticated engineering tools for creating, analyzing, and testing different speech sounds. They did so in the hope of identifying the basic building blocks of speech perception and production, and they, too, limited themselves to the characterization of a language's inventory of sounds, which stopped short of being a generalized theory that holds of all human speech.

Behaviorists, led by psychologist B. F. Skinner (and supported by prominent philosophers such as W. V. O. Quine), sought to study language behavior but explicitly barred themselves from a scientific inquiry into the cognitive mechanisms that underlie this behavior. Alluding to limitations imposed on "sound" scientific inquiry, Skinner argued that scientists cannot posit abstract cognitive mechanisms as "inner causes" of behavior, because these are not directly observable; therefore, scientists can only make connections between observed environmental factors that the organism experiences (a.k.a. stimulus) and the behavior it displays subsequent to this experience (a.k.a. response).[6] The rejection of grammatical concepts was quite natural, as they are not directly observable, and hence are devoid of scientific value, according to this highly influential stream in psychology. Grammar, Skinner claimed at the heyday of behaviorist domination of American psychology, should be barred from use in scientific inquiry.[7]

Influenced by behaviorism, students of language turned to human behavior in the language domain, which they sought to study *statistically*, in a manner that excluded grammar. They took a slightly different path, though: attempting to develop tools for quantitative investigations of linguistic behavior, they turned to the recently invented information theory. This approach offered mathematical tools for the measurement of informational content in transmitted messages, quantitatively expressed in terms of the amount of uncertainty that is removed when a new piece of information becomes input.[8]

When applied to human language, this approach was only concerned with the quantification of the degree of uncertainty at a particular serial position in an incoming string. Psychologist George Miller analyzed human language behavior with these tools.[9] He reasoned that when we listen to a sentence, we typically try to predict the next word from the already-heard context. For example, *After pouring milk into her empty bowl at breakfast, Jeannie realized that an ingredient was missing and ran to the convenience store to get* ________. Given the preceding context, I believe that most Americans would vote for *cereal* (or perhaps *muesli* if your heritage is German or Swiss) as the most likely candidate next word (some might instead say *her*, thinking that the continuation would be *her morning cereal*). Given this expectation, the reduction of uncertainty would be small upon encountering the word *cereal*. If instead the next word were surprising (e.g., *poison*), the reduction of uncertainty would be great. Using this logic, Miller investigated how well people predict the next word, and he tried to model his results with information-theoretic tools. As no assumption of unobservable "inner causes" was required (especially grammar, hierarchy, parts of speech, or grammatical relations among parts of a string), Miller's approach gained popularity among behaviorists.

Finally, there were the *culturalists*, whose interests contrasted sharply with those of the other groups. The culturalists viewed language as a window into culture, and they were therefore focused on differences between languages and their connection to cultural variation. Best known among these linguists are Edward Sapir and Benjamin Lee Whorf,[10] who embarked on expeditions to small Indigenous communities whose members spoke languages like Inuit or Hopi, in the hope that documenting these languages would enhance our understanding of the relation between language and culture. Thus, the absence of past tense inflection in Hopi indicated to Whorf that the concept of time in Hopi culture was different from the European concept, a difference he tried to investigate.

However, this (mostly anthropological) linguistic prong did not apply a uniform system of grammatical description. It moreover remained focused on differences between languages at the single word level and on how these differences could serve as windows into the particular cultures in which these languages were used. In contrast to the previous approaches, culturalism did not look at human behavior at the individual level but rather took a broad look at the language–culture relation. Its connection to the present discussion is therefore rather tenuous.

Birth of Generative Grammar

Against this backdrop, Chomsky developed a new method of linguistic analysis, which mostly focused on syntax: the analysis of sentences into their constituent parts. Criticizing the above approaches, Chomsky realized that the simple principles can at times be overly simplistic.[11] Take linear proximity, for example: When you ask a question like *Can birds that fly run away?*, a *yes* answer is appropriate if they can run away, but not if they can fly or are flying. Why does the verb *can* relate not to its closest ally, *fly*, but rather to a distant one, *run away*? Chomsky showed how his work—but not others'—offered an answer (that the relevant notion of distance is not linear but structural: it takes hierarchical relations into consideration). Like his teacher Harris and the other structuralists, he derived immediate constituents by rules that annotated the diagrams with the familiar grammatical categories: Noun (N), Verb (V), Adjective (A) Preposition (P), Sentence (S), and some others. Using these basic building blocks, he formalized a system of *phrase structure rules*. This system used a finite set of rules, symbols, and elements to project into an infinite set of strings. He also added new methods and showed how they were applied in the analysis of phrases and sentences. He also designed empirical tests, to justify the notion of constituent. Let's see some examples, so you can get a flavor of this early rule system.

Chomsky posited grouping rules, whose general format was X→Y+Z, where X, Y, Z are categories like Sentence, Noun Phrase, Verb Phrase, Noun, Verb, and so on. These rules work through rewriting, designated by an arrow (→). You rewrite the single category that is to the left of the arrow (X)—starting with the top category, Sentence—as the categories to its right (Y, Z): Sentence→NP+VP (or S→NP+VP for short). You continue with intermediate rules that derives smaller pieces (e.g., VP→V+ NP), and you continue until you get all the way down to the words, which are derived via "category labeling rules," for example, N→*you* and V→*flew*. Thus, to analyze the simple sentence *You flew*, you apply S→NP+VP; NP→N; VP→V; and finally, the category labeling rules.

Word strings that cannot be analyzed with this given set of rules, for example, **Fly birds*, are deemed ungrammatical, because the tiny system I just sketched features no rule that enables the verb to precede its subject noun. We can therefore evaluate the correctness of a grammar by testing whether or not strings it is capable of analyzing are grammatical by our intuitions—by asking ourselves (or others) whether the line between good and bad sentences that the grammar draws suits our own judgments.

One immediate advantage of this system, which Chomsky was quick to point out, is that it can accommodate structural ambiguity. Certain phrases and sentences can be analyzed in more than one way, with obvious consequences for meaning: the phrase *the first parliament member* may be analyzed as *[the [first* ***[****parliament member****]***]] and end up meaning 'the first among many members of parliament' or as [the [***[****first parliament****]*** *member*]], with a meaning 'the member of the first parliament (ever).' Chomsky's famous example was the ambiguous sentence *They are flying planes*. He used the rules to assign it two different syntactic analyses, giving a formal expression to the observed ambiguity. Here are the rules, which can be applied in two different ways, to result in two graphical forms (tree structures) in figure 1.2:

Sentence → NP + VP
VP → V+ NP
NP → *they*
NP → *planes*
V → *are*
NP → *flying planes*
V → *are flying*

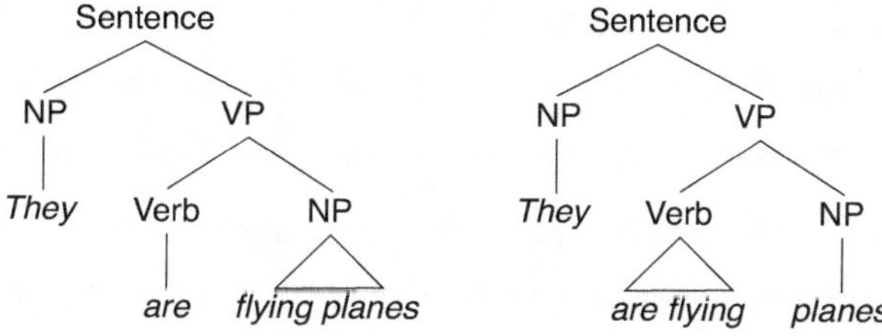

Figure 1.2
Two different labeled trees of a structurally ambiguous sentence (Δ marks a part whose internal structure is suppressed). Each labeled tree gives rise to a different meaning. Left: 'They are the planes that are flying.' Right: 'They are the persons who are flying the planes.' A playful reader may enjoy keeping track of the rules applied at each level of the two trees.

To Chomsky, we humans must possess a cognitive system that can create such trees, or else we would never be able to detect the ambiguity. Still, the example sentences thus far considered have been relatively short and simple, which means that there is more to human linguistic cognition than these examples illustrate—we and our grammar can generate and analyze arbitrarily long sentences. To illustrate this, Chomsky highlighted the grammar's device of *embedding*. This important grammatical tool that we have at our disposal enables us to nest one sentence within another (e.g., *John said that Mary loves cookies*)—and then to further embed yet another sentence (e.g., *John said that Bill thought that Mary loves cookies*) and keep doing so, perhaps ad infinitum. This is a most powerful ability. Indeed, it is widely agreed today that our cognitive syntax has the property of recursiveness.[12] The rule system we build must therefore capture it. To this end, it must be endowed with a new power: it has to be recursive.

Embedding and Recursiveness in Syntax: An Example

We use the grammar's embedding tools many times a day, nesting one sentence within another. Below, I illustrate how a tiny ad hoc rule system ("a toy grammar") generates embedded sentences. The previous rule system has been augmented with one category; let's call it Sentence Glue (SG) for illustrative purposes. This category enables a speaker to glue, or embed, one sentence within another, an operation that may recur any number of times. The tree structure in figure 1.3, associated with the following rules, shows that the top rule (Sentence → NP + VP) was applied three times in the sentence *John said that Bill thought that Mary loves cookies.*

Sentence → NP + VP
SG → *that* + Sentence
VP → V+ SG
VP → V
NP → *John, Bill, Mary, cookies*
V → *said, thought, loves*

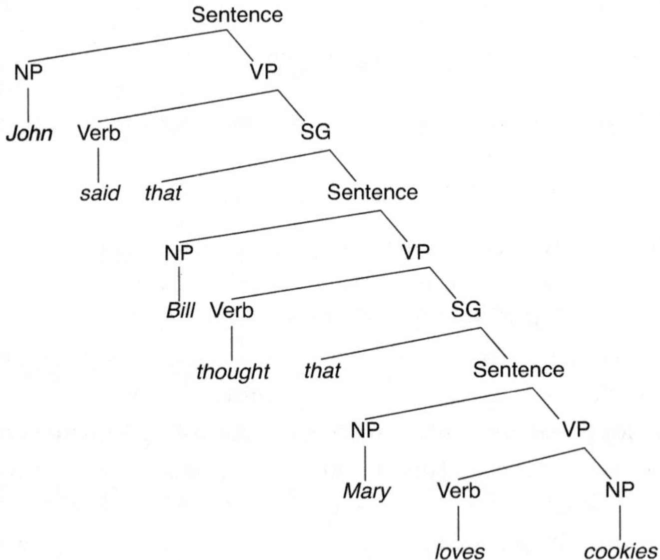

Figure 1.3
A sentence with two embeddings. The category Sentence is called twice by the rule SG → *that* + Sentence, which allows for three occurrences of Sentence, each expanded three times into NP + VP.

A recursive function is one that can call itself (whether directly or indirectly). For the example just given, this system must have rules that allow it to analyze a sentence within another sentence. To equip our phrase structure rule system with this property, we need to modify it a bit. Thus far, the category to the left of the rule's arrow was rewritten as two others to the right. Take in particular S → NP + VP. The rules NP → N and VP → V + NP have no way of calling S again. But to embed a sentence within a sentence, we need to allow rules that call S. In other words, S should be allowed to be on both the left and right sides of the arrow (e.g., S → NP + S); we can also achieve the same effect with rules that allow us to get from S to another S in two steps (e.g., S → NP + VP; NP → N + S). Note that we can now keep expanding our rules on end: S → NP + VP; NP → N + S; S → NP + VP; NP → N + S . . .). This gives our rule system a lot of power: we can embed time and again.

Early TGG and Its Properties

Chomsky then moved on to other important relations between sentences, and he set up rules to account for them. He highlighted the intimate relationship between the active sentence *The man ate the food* and the sentence *The food was eaten by the man*, which contains roughly the same words and has roughly the same meaning but is in passive voice. This, he argued, is an important relation that the grammar should not miss. He then showed that his phrase structure rules, which assumed linear adjacency between constituents, were incapable of capturing the relation between these two sentences: to get the passive from the active, you need to switch the positions of *the man* and *the food* in the sequence, then change the verb *ate* to the participle *eaten*, and finally, insert the auxiliary verb *was* and the preposition *by*. With a grammar that only has phrase structure rules, you cannot carry out such operations.

Chomsky's new grammatical system was designed to solve such problems. First presented in detail in his 1957 monograph *Syntactic Structures*,[13] it contained two components. First was a system akin to Harris's phrase structure rules, which analyzed, or generated, sentences that were annotated for their immediate constituents. Then, to fix the active–passive problem described above, this rule system was augmented with a system of *grammatical transformations*, rules that have a different shape: given the sequence of words in the sentence (a concatenation), as generated by the phrase structure rules, they may switch the sequential order, substitute some words, and add or delete others.

I suggest taking a minute to see the fine mechanics of this 1957 system through the active–passive sentence pair *The man ate the food* and *The food was eaten by the man*. The sequence of operations takes the active sentence *The man ate the food* and first applies the rule $S \rightarrow NP + VP$. This operation yields $NP_1 =$ *[the man]*; $VP =$ *[ate the food]*. The next step parses the pieces into yet smaller ones: $VP \rightarrow V + NP$ results in $V =$ *ate*; $NP_2 =$ *[the food]*. Finally, the verb is broken into a tense marker (categorized as an Auxiliary, or Aux for short) and a verbal stem: $V \rightarrow Aux + V$, results in *ate* = [past]*[eat]*. The sequence of labels of the resulting string is $NP_1 + Aux + V + NP_2$.

This string is then input to the passive transformation: it is on the left side of the double arrow symbol ⇒ below. On the right side is the output of the transformation: the sequence of categories and the special words of the passive sentence.

$$NP_1 + Aux + V + NP_2 \Rightarrow NP_2 + Aux + V + \textit{en by } NP_1$$

Expanding the categories to their terminal nodes, we obtain *The man ate the food* ⇒ *The food was eaten by the man*.

This setup has several implications: the grammar is now more complex, because it is divided into two rule types: the first type features generative "structure-building" phrase structure rules; the second is a new rule type—a grammatical transformation, which relates pairs of sentence types (such as active–passive, question–answer, and much more) to one another. Thus, phrase structure rules generate basic structures, and grammatical transformations operate on the basic structures that are the output of these rules/transformations and thus account for regular relations among structures. The result: transformational generative grammar (TGG), as it became known. Transformations enrich a relatively limited inventory of basic (or "kernel") sentences of the language: they enable derivation of passive sentences, questions, and other complex sentences, which are just one or two transformations away from their active declarative counterparts. Transformations constitute a purely syntactic operation, one that (supposedly) does not affect meaning. The grammar becomes more cumbersome, but the payoff is an explanation for a new set of phenomena: we get an answer to the question why languages feature pairs of sentences that have the same words and the same meaning, despite the fact that they differ in their syntactic form. The answer should be clear by now: one member of the pair is derived by a syntactic transformation from the other. Note that words in this grammar are syntactic objects, an integral part of the phrase structure rule system (e.g., there may be many rules of the shape V→*eat*). This would later change: a reformulated grammar would accommodate a separate store of words, or mental lexicon, with its own properties.

With a system of rules at hand, Chomsky proceeded to argue that syntax is blind to meaning (let alone connotation): a well-known example he gave contrasted two nonsensical word strings, *Colorless green ideas sleep furiously* and *Furiously green sleep ideas colorless*. Both, he observed, make no sense, and yet only the former is grammatically well-formed. Syntax, he concluded, is therefore about form and is independent of meaning and context. He thus embarked on the development of methods for the study of syntactic structure that is independent of content.

How does this affect alternative approaches to grammar? The cultural approach immediately becomes unrelated to Chomsky's object of inquiry. As for the statistical approach, he criticized it by showing that an information-theoretic system, incapable of analyzing sentences into a hierarchy of parts and subparts, misses important linguistic generalizations. An example he gave was the complex sentence ***If*** *[John sleeps late]* ***then*** *[he will miss the train]*. Its logical and syntactic structure, he argued, resembles ***If*** *[I don't eat the carrot],* ***then*** *[I can't lift the weight]*: while apparently unrelated, both sentences conform to the conditional structure ***If*** *Sentence*$_1$ ***then*** *Sentence*$_2$. But if we look at the neighborhood relations of the words ***if*** and ***then*** in these two sentences, we get a different perspective: the first ***if*** has a sleeping context, whereas the second, an eating one; the first ***then*** has a train-missing context, whereas the second has a strength one. Chomsky wrote that if we start just by looking at neighborhood relations and count how many times ***if*** and ***then*** appear in particular contexts, we would be led to treating these two sentences as unrelated, and we would miss an important generalization: that the two sentences have common structural properties and participate in the logical relation ***if . . . then . . .*** To Chomsky, this was among many profound and critical regularities in human language that a purely statistical perspective misses, which indicated that it was deeply lacking.

This is how TGG began. Importantly, in making his then-novel proposals and contrasting them with extant alternatives, Chomsky followed a methodological strategy, one philosophically known as "inference to the best explanation." The account he proposed was not only supported by evidence; it was also accompanied with a demonstration that it was superior to alternative accounts (structural, statistical, cultural), having solved problems that they had not.

Much has been written on the skepticism, to say the least, with which this new approach was greeted. I will only note that traditional linguists argued that it only focused on English, ignoring other languages; philosophers accused it of assuming an obscure version of mentalism, even dualism (by which minds are not causally related to bodies); psychologists chided its bold acceptance of inner, unobservable, cognitive processes. But even its fiercest critics today would admit that it heralded the cognitive revolution, which placed its focus on the mental, and then neural, mechanisms that govern human behavior, rather than on behavior itself. Some of these issues will play a role in chapters to come.

With this short history in our pocket, we can fast-forward to today. Whereas the legacy of this 70-year-old system does live on in certain respects, its descendants look rather different. While there are at present multiple versions of linguistic theory, which debate one another, they all seem to be more generalized, more constrained, more precise, and more abstract.[14] Possibly, many if not most of them are even more explanatory than their 1950s predecessor, and their empirical scope is certainly much broader. Let's look into this in some detail.

What the Linguist Is Talking About—The Concepts "Grammar" and "Grammaticality"

Grammatical transformations, as a mere technical innovation, do not a scientific revolution make. Indeed, the new linguistic theory in which Chomsky couched his technical ideas had larger scientific goals. In a nutshell, he looked beyond the descriptive horizon of his predecessors, viewing linguistics as being not about linguistic description but rather about the *human knowledge* codified in this description. In other words, TGG, if accurate, properly concise, and comprehensive, is not a mere grammar book for foreign language learners.

So, what is the nature of this construct, TGG, that we are said to possess deep inside our minds (and brains)? It is a theoretical account of what human language users know about their language—knowledge by virtue of which they are capable of using language the way they do. Importantly, this theory is not just about what humans do when they use language; it is about what they *can* do with language, as owners of a human language faculty. The point pertains to all grammar rules, from phrase structure rules ($S \rightarrow NP+VP$; $NP \rightarrow N+S$; $VP \rightarrow V+NP, \ldots$) to the complex transformational rules ($NP_1+Aux+V+NP_2 \Rightarrow NP_2+Aux+V+en$ *by* NP_1) down to that little *-s,-z,-ez* rule that every native speaker of English knows. The last rule, in fact, is followed strictly, excluding even wrong plurals of invented words, as a famous experiment showed long ago. Children were asked to turn a singular noun into its plural form. They were shown an unfamiliar object and told, "This is a *wug*!" Then, they were shown two such objects and asked, "What are these?" Remarkably, even preschoolers returned *wug-z*, not **wug-s* or **wug-ez*.[15] Speakers seem to have clear intuitions about grammatical well-formedness at the word, phrase, sentence, and text levels. Notably, these intuitions are rather uniform across native speakers of the same language (there are some sources of noise and distortion, which we linguists have learned how to control and filter out).

The set of rules that the linguist posits is therefore expected to cover all the well-formed sentences in a language that a native speaker may ever utter, and only these. The actual status of sentences (distinguishing between, e.g., the well-formed *The man ate the food* and the ill-formed *The man ate a food* and *The food ate the man* and certainly *Food the ate man the*) is determined solely by the testimony, or judgment, of native speakers of the language—evidence obtained by behavioral experiments.

If TGG is capable of distinguishing between the well- and ill-formed sets of word strings and if it is the best way to do so that is found on the market of ideas, then we must conclude that *it is* the very knowledge that the speaker possesses. Such a bold claim is again in keeping with the well-known philosophical principle "inference to the best explanation" (also known as abduction), which seems to be followed in most sciences: if you have several accounts of a set of natural phenomena and if you also have criteria that help you decide which account is best, then it follows that the best explanation is the *true* theory, until further notice.[16]

Pursuing this line of scientific reasoning, Chomsky's research program aimed to characterize the human grammar, which should lead to an understanding of the human communicative ability—our exceptional ability (perhaps unique within the animal kingdom) to produce and comprehend any possible utterance in our native language, regardless of whether we have previously heard or read it.

Are We Born with a Grammar in Our Head? The Innateness Hypothesis

If speakers of English know its grammar (and that includes young preschoolers, who say *bee-z* and not *bee-s* just like everyone else), how did they get to know it? Did the rules get into their heads by rote learning? Was there a teacher? Moreover, if English speakers know their grammar, why can't they just write down its rules?

An influential idea for an answer was articulated by Plato some two thousand years ago, in his dialogue *Meno*. The issue debated there is whether knowledge is acquired through the recall of preexisting ideas or through learning by exposure to information (a.k.a. the nature vs. nurture debate). Through a Q&A experiment with a slave boy, Plato argues for the former position. That is, he claims that much of the knowledge we possess is actually ours from birth and only needs triggering via an interaction with the environment. Some knowledge, then, is innate. Chomsky, relying mostly on Descartes's later endorsement of a version of this idea, argued for innate knowledge from a linguistic angle in his celebrated *poverty of the stimulus* argument. We have an ability to produce and comprehend grammatically well-formed sentences that we have never seen, he said. This ability, to generalize from examples to a rule, is not learned. Hence, aspects of our grammatical knowledge transcend our past experience and must be innate.

Chomsky's fresh approach and sharp arguments, as well as his personal charisma, won him young supporters early on. As philosopher John Searle put it, "Chomsky did not convince the established leaders of the field, but he did something more important, he convinced their graduate students."[17] George Miller, the mathematical psychologist who had started out with statistical models for next word prediction, was among the first to change his views, and he become an early collaborator. Together, Chomsky and Miller wrote a program for the new language science, which consisted of three questions:

A. What is the precise nature of linguistic ability?

B. How does it arise in the individual?

C. How is it put to use?[18]

At the time, little was known about the biological bases of linguistic ability. Later, two additional foundational questions were added to the list:

D. How does the human brain support this ability?

E. Is it uniquely human and how did it evolve?

It appears, then, that the new science of linguistics has touched on perennial psychological and philosophical questions from a linguistic angle; later, issues concerning its neural bases also arose. It is for this reason that this science has had far-reaching implications; it is also one of the reasons why it ignited so many debates and controversies and has acquired a considerable number of critics and enemies.

The Precise Nature of Our Linguistic Ability

Seventy years past these early efforts, the current theory is obviously different. Generations of linguists have improved it vastly, broadened its empirical coverage, and turned it into a highly formal and constrained theoretical apparatus. Still, the ultimate goals of this theory have largely remained similar to those Chomsky originally formulated: to account for human linguistic ability.

What, then, is the nature of this ability and how does science characterize it? If linguistic ability were only characterized by the sentences humans produce in speaking and writing (perhaps with special emphasis on those that are most frequent, like *The man ate the cake*), the linguist's life would be relatively simple. Almost anyone who has taken a basic computer science course could write a grammar for such a language, which would likely implement some basic phrase structure rules, as was done above. That should not be too difficult for a programmer—after all, Chomsky's early work laid the foundations of the theory of programming languages and their complexity.

The matter becomes more serious once properties of language that are less visible are taken into account. We saw how the earliest version of TGG rationalized the need for transformations by observing abstract properties of the language: the theory created a typology of sentences (active–passive, question–answer, and more) and argued that these types needed to be related to one another, which was the function of the transformational rules. The relation between such sentence pairs is an abstract property of language, which must be accounted for. Transformational rules, however, complicate the grammar. Indeed, the 1950s syntax featured two distinct rule types, with a clear ordering imposed on their application. Phrase structure rules were used to build structure, and the output of this process was then fed into transformational rules. Take active and passive sentences. The thought was that, while they may sound different on the surface, deep down their meanings are identical. By the 1960s, this thought was made explicit: pretransformational representations were called "deep" structures, whereas their posttransformational derivations were called "surface" structures. Active and passive sentences thus had the same "deep" structure but differed in their "surface" structures—the external shape of the sentence as we finally utter or hear it.

The Cognitive Architecture of TGG—Take One

Each sentence had two representations: "deep" and "surface." For an active sentence, they contained the same word sequences, but for a passive sentence, different ones: the deep structure of the passive fed into the transformational rule $NP_1+Aux+V+NP_2 \Rightarrow NP_2+Aux+V+en+by+NP_1$, making its overall structure more complex than its active counterpart. To simplify matters, we can choose the total number of symbols of the rule as our complexity metric (repetitions allowed); there are seven symbols to the left of the arrow (NP_1, +, Aux, +, V, +, NP_2) and 11 to the arrow's right (NP_2, +, Aux, +, V, +, *en*, +, *by*, +, NP_1). The grammatical system of TGG was shaped as in figure 1.4.

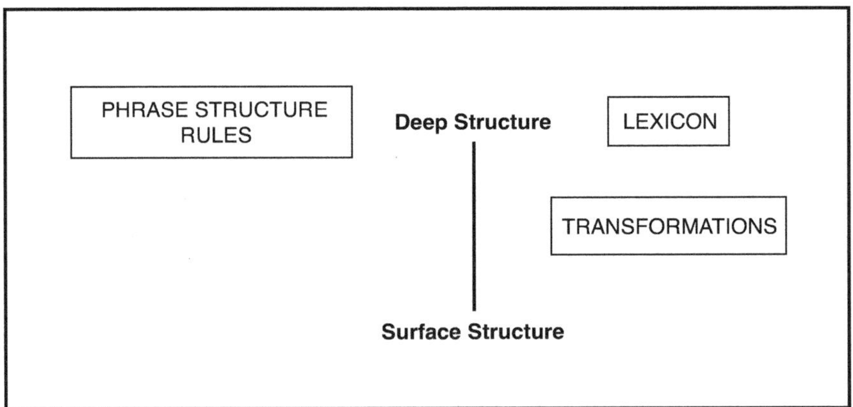

Figure 1.4
The structure of the grammar, Take One. Each sentence is associated with two representations. Deep Structure is fed by phrase structure rules that combine with words pulled out of the lexicon. The result is mapped onto Surface Structure by transformational rules (which may or may not apply).

On Take One, TGG posited syntax as a specialized cognitive system for syntax that interfaces with the semantic system and with general cognition via Deep Structure. Deep structures are stored in every speaker's brain. At its other end, Take One had Surface Structure, interfacing with the sound system of language. Transformations mapped Deep Structure onto Surface Structure and vice versa. The path from meaning to sound and back was assumed to be through the syntax, which, upon investigation, was becoming increasingly complex.

The metaphors "deep" and "surface" inspired many thinkers and aggravated some, leading to multiple attacks on the theory, although the representations were neither deep nor surface—these names merely denoted abstract objects, the grammatical descriptions of sentence structure.[19]

This is how the nature of grammar was viewed. It was further seen as a key to our mental makeup, as it enabled rich and flexible communication. In the next chapter, we dig a bit deeper into its later development, which I dub Take Two.

Taking Stock

In this chapter we saw:

- Snippets of historical grammars, from Pāṇini's Sanskrit grammar
- Three twentieth century approaches to linguistics
- Reflections on what grammar might be
- Structural ambiguities providing motivation for immediate constituents
- Chomsky's phrase structure rules as a basic way to generate structure. Their recursive property enables embedding (in fact, as many embeddings as you wish)
- Chomsky's transformational rules—part of TGG—accounting for regular relations among sentence types, such as active–passive and question–answer
- The broader significance that TGG has for psychology, philosophy, and neuroscience
- Chomsky and Miller's program for the language sciences
- Innate ideas then and now (from Plato to Chomsky)
- Take One of the architecture of our cognitive syntax: Deep and Surface Structure

2 Hints from Present-Day Linguistic Theory

Syntax, my lad. It has been restored to the highest place in the republic.
—John Steinbeck on JFK's inauguration speech, 1961

You know as much about this as I do. Argue with me! Teach me something I don't know!
—Morris Halle, cofounder of the MIT Department of Linguistics

The Cognitive Theory of Syntax

We now turn to the development of the cognitive theory of syntax, not before acknowledging that the short introduction in the previous chapter left many general issues untouched. My intention here is not to teach you linguistics but only to introduce the main problems the field is trying to solve, the rationale of the enterprise, and some basic theoretical ideas and tools that are common to virtually all theoretical frameworks that are currently on the market. Finally, I discuss solutions that linguists offer to certain problems, along with implications these solutions may have for philosophy, psychology, and neuroscience. As noted, once matters get a bit technical, the text is boxed and marked gray. You may skip these parts, mostly with impunity (though you'll be missing out on some fun stuff).

A central theme that I leave out is cross-linguistic variability. The legend of the Tower of Babel may have no historical basis, but the fact is that at present, there may be as many as 7,000 languages spoken in the world. These may have different grammars, and TGG as thus far presented obviously cannot be the grammar of all of them. Briefly, the vision (major parts of which have actually materialized) is to have a unified and abstract theory of syntax, with built-in parameters that allow for limited variation across languages. To take the phrase structure rules as an example, they work for English but not German. In English you say *I have* ***seen*** *the girl,* the verb *seen* appearing before the object, whereas in German, it is at the end: *Ich habe das Mädchen* ***gesehen.*** English and German are thus similar in that every sentence has a subject and a verb, potentially an object, and so on—but the grammars also appear to be different; the examples just given show that the position of the main verb is not the same. This difference forces the formulation of rules that are sufficiently abstract and tight but also sufficiently flexible to accommodate such similarities and differences.

While cross-linguistic issues occupy the minds of many, they go beyond the present scope. I will only note (bringing in no evidence) that there is a level of description (sometimes called Universal Grammar) at which the shape of all human grammars, though not the specific rules themselves, is one and the same. Languages and their grammars do differ, but the cognitive principles behind them do not, as far as we know.

Questions and Answers—Critical for Linguistic Ability and Functioning and for Linguistic Theory

As repeatedly noted, you can't do everything in one book. The topics discussed here have been chosen for their relevance, as well as for both historical and illustrative purposes; they are intended to underscore difficult but central empirical problems for the theory of linguistic cognition and to put on display theoretical efforts to understand them and creative moves made to help in that endeavor.

For some philosophers, language may exist in the absence of communication between individual agents. They view it as an internal cognitive tool for thought rather than communication.[1] But even these thinkers agree that subsequent to its evolutionary development, this tool was externalized (or "exapted")[2] to serve as a tool for communication or information exchange among agents. A good place to start our discussion would therefore be a paradigmatic type of communication scenario, namely, requests for information and informative responses to these requests—in our case, English questions and their answers:

1. Q: **Who** ate the chicken? A: **John** ate the chicken.
2. Q: **What** did John eat? A: John ate **the chicken.**

These are good Q–A pairs under normal circumstances, given cooperative and truthful interlocutors with a shared body of contextual knowledge (in this case, that there was more than one potential eater; that John is that guy over there; that among the food items around, there was a single chicken). It is of note that such exchanges are not always successful. Schizophrenic patients famously answer questions "tangentially," responding to them, sharing information, but without answering. For example, Q: "How have you been feeling today?" A: "My flat is just around the corner." Here, I discuss the grammatical properties of *successful* exchanges via Q–A pairs.

English question words are typically at the beginning of a question sentence (as *who* and *what* are in the questions in 1 and 2), while the center of the answer (the bold element in the answers in 1 and 2) may be in another position. In 1, the question word and the center of the answer are both in a sentence-initial position (this type of question is known as a subject question). In 2, they are not: the question word(s) is sentence-initial as before, but the center of the answer is sentence-final (an object question). In the question in 2, moreover, the verb *ate* breaks into two nonadjacent pieces, the past-carrying *did* and the verb *eat* (reminiscent of the treatment of the verb by the passive transformation; see chapter 1).

Another oddity in 1 and 2 is that while the two questions differ from each other, the shape of the answers is the same—and yet they do not convey the same information. A question carries a request for an unknown piece of information; a verbal answer to the question is expected to convey that information, usually through a declarative sentence. It is but natural to think about this unknown as a variable, whose value will depend on the question's content and the context in which it is asked. We can reformulate the Q–A pairs in 1 and 2 in a form that is a bit logic-like in flavor:

1′. Q: *What is the identity of* ***x****, such that* ***x*** ate the chicken?

A: ***x*** **= John** (ate the chicken)

2′. Q: *What is the identity of* ***x****, such that* John ate ***x***?

A: (John ate) ***x*** **= the chicken**

Speakers are of course aware of these properties of Q–A pairs (even if we may not always be able to articulate them properly). We ask and answer questions daily with little difficulty; the theory of our internal grammar must explain how such exchanges work seamlessly and why deviations from the norm (like those in schizophrenia) are so rapidly detected. So, there are two goals here. First, to set up a rule such that "deep" structures of Q–A pairs are closely related (similar to active–passive pairs, as discussed in chapter 1). Second, to design a syntax for 1 and 2 that leads to the "meanings" in 1′ and 2′. For simplicity, we will let ourselves be slightly sloppy: for ease of reading, let's suppress the phrase structure and category notation.

A suitable TGG question–answer rule may be a syntactic transformation that establishes a relation between a *who* or *what* question and its answer by (*a*) choosing a noun phrase from the answer (*John, the chicken*) and replacing it with the proper question expression containing the variable ***x*** (the italicized string in 1′ and 2′); (*b*) copying this expression to the front of the sentence; (*c*) replacing the original copy with the variable ***x***. Take a minute, and try to apply these three operations to the A part of 1 and 2; you should get the Q part of 1′ and 2′, respectively. This question transformation expresses the syntactic relation between an answer and its question. In the spirit of Chomsky's early work, here are possible transformational rules (with some abbreviations, for ease of reading, e.g. the auxiliary *did* in 2 is suppressed in the notation). On the left side of the double arrow ⇒, there is a structural description of the sentences that would undergo the change, described on the arrow's right side.

1″. **Subject question**

A: $NP_1 + V + NP_2 \Rightarrow$ Q: ***what x.x*** $+ V + NP_2$

2″. **Object question**

A: $NP_1 + V + NP_2 \Rightarrow$ Q: ***what x.***$NP_1 + V +$ ***x***

Note that this rule is formulated backwards (from A to Q) and that it (intentionally) emphasizes the question–answer relation. The representations in 1″ and 2″ add a semantic perspective: the question operator ***what x*** (which stands for a request to convey *the identity of* ***x***) and the variable ***x*** indicate that *who* and *what* questions are requests for information about an unknown. The location of ***x*** marks the position of this unknown entity in the question sequence. This variable is connected (by a logical relation called binding) to its position in the answer.

This complex representation contains hints about the complexity of the communication process. Yet this complexity does not hinder communication. Q–A exchanges occur frequently and without much reflection—we seem to know what we are doing, or else communication would constantly break down.

How Far Can a Question Go?

What is the difference between the Q part in 1 (Q_1) and the Q part in 2 (Q_2)? Both intuition and experimental evidence suggest that perceptually, Q_2 is more difficult to understand than Q_1. This may be due to two properties of their respective representations. Perhaps somewhat naïvely, it appears that, as Q_1 is a subject question, all one needs to do in order to convert the answer into a question is replace John with ***what x.x***; whereas in Q_2, an object question, there seems to be an added displacement as the question expression—***what x***, a.k.a. *What is the identity of **x***—is copied to the front. Relatedly, the linear distance between the question expression and ***x*** is larger in Q_2 than in Q_1.

We will delve into the reasons for the difficulty differential later (in the context of the syntactic brain). At present, we focus on the observation that questions may displace a constituent at various distances. How far can ***who***/***what*** travel (a factual issue), and will it affect the question transformation in 1″ and 2″ (a theoretical issue)?

Facts first. Verbs like *think* may help us track distances between the relevant positions of the words (or phrases) under consideration. Such verbs typically are followed by (or more accurately, embed) a sentence that describes the content of a thought:

3. Q: ***Who*** did *Mary think* ate the chicken? A: *Mary thought* **John** ate the chicken.
4. Q: ***What*** did *Mary think* John ate? A: *Mary thought* John ate **the chicken.**

Applying our quasi-formal rewording routine, we get:

3′. Q: *What is the identity of **x**,* such that *Mary thinks **x*** ate the chicken?
A: ***x*** =**John** (Mary thinks ate the chicken)

4′. Q: *What is the identity of **x**,* such that *Mary thinks* John ate ***x***?
A: ***x*** =**the chicken** (Mary thinks John ate)

Embedding made the sentence longer and perhaps more complicated, yet our method allowed us to express the speaker's intention in this way. However, as the reader may verify, the transformational rules in 1″ and 2″ no longer apply: the current string of words does not match the transformation's structural description (on the left side of the arrow). To fix that, we could try to add ingredients to the transformation. It would become more cumbersome, but it would fit the new sentences (again, the auxiliary *did* is suppressed for simplicity):

3″. **Subject question**$_{think}$

$NP_1 + V_1 + NP_2 + V_2 + NP_3 \Rightarrow$ ***Who/what x***.$NP_1 + V_1 + \boldsymbol{x} + V_2 + NP_2$

4″. **Object question**$_{think}$

$NP_1 + V_1 + NP_2 + V_2 + NP_3 \Rightarrow$ ***Who/what x***.$NP_1 + V_1 + NP_2 + V_2 + \boldsymbol{x}$

This move, which adds special rules designed to accommodate this specific embedding under *think*, is not only heavy-handed and ad hoc; it is also hopeless, because language allows the freedom to embed multiple times and still ask questions and give answers, whereas the rules in 3″ and 4″ do not. Keeping with the rule-adding strategy would force us to formulate a special rule for each embedding, which would only blur our vision, not sharpen it, as it would make the general property of the transformation evaporate into a multitude of specific rules:

5. **Subject question**$_{think\text{–}believe}$

   Q: ***Who*** did Mary think *Bill believes* ate the chicken?

   A: Mary thought *Bill believes* **John** ate the chicken.

6. **Object question**$_{think\text{–}believe}$

   Q: ***What*** did Mary think *Bill believes* John ate?

   A: Mary thought *Bill believes* John ate **the chicken.**

Questions can be asked from the bottom of even longer sentences, with three embedding verbs:

7. **Subject question**$_{think\text{–}say\text{–}believe}$

   Q: ***Who*** did Mary think *Sue said Bill believes* ate the chicken?

   A: Mary thought *Sue said Bill believes* **John** ate the chicken.

8. **Object question**$_{think\text{–}say\text{–}believe}$

   Q: ***What*** did Mary think *Sue said Bill believes* John ate?

   A: Mary thought *Sue said Bill believes* John ate **the chicken.**

Something more abstract and general is needed, in order to precisely codify the question expressions' apparent freedom to travel large distances, across many nouns and verbs.

A promising possibility leaves the context blank and allows the transformation to create questions everywhere, as long as they position the question expression on the left edge of the representation:

9. **Subject question**

   $\ldots \text{NP}_n + \text{V}_n + \text{NP}_{n+1} \Rightarrow$ ***Who/what x*** $\ldots \boldsymbol{x} + \text{V}_n + \text{NP}_{n+1}$

10. **Object question**

   $\ldots \text{NP}_n + \text{V}_n + \text{NP}_{n+1} \Rightarrow$ ***Who/what x*** $\ldots \text{NP}_n + \text{V}_n + \boldsymbol{x}$

The symbol "..." stands for "any otherwise legitimate sequence of words." This move would force a new rule type—one that contains a variable that can take any legitimate context as a value. This way, the transformations in 9 and 10 would account for 5–8 (e.g., for 5, just replace "..." of the rule with "*did Mary think Bill believes*" and you will see that the rule works!).

This more relaxed formulation of the transformation seems to work. But new problems arise as the rules are modified from the rigid but unrealistic 3″ and 4″ to the somewhat looser 9 and 10. The latter harbor incorrect predictions. The Qs in 11 and 12 are in keeping with these rules (for 11, replace "..." of the rule with "*the woman* **who** *taught*"; for 12, replace "..." with "*the thug* **who**"), and the linear distance between **who** and its original position is shorter than in the case of ***what*** in 8, but the Qs are nonetheless ungrammatical (marked "*" and annotated henceforth as Q*). Importantly, the request for information each of these bad questions conveys makes perfect sense, as seen by the paraphrases Q✓. Yet, as stated, no Q can be asked.

11. A: Bill married the woman who taught **his sister.**

    Q*: ****Who*** did Bill marry the woman ***who*** taught ~~who~~?

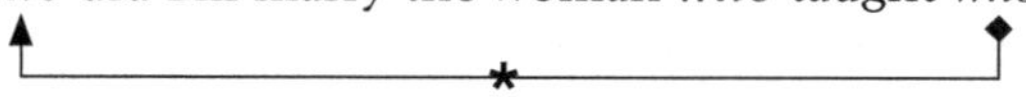

    Q✓: ***Who*** is the person such that Bill married the woman who taught her?

12. A: The thug who **Bill** saw was scary.

    Q*: ****Who*** did the thug ***who*** ~~who~~ saw was scary?

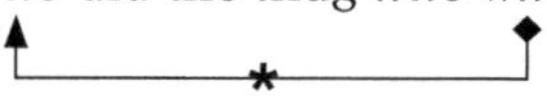

    Q✓: ***Who*** is the person such that he saw the thug who was scary?

I have added two helpful notational innovations: an arrow marks the link between the target and original positions—the sentence-initial copy of ***who*** and the original ***who*** (the one that replaced the boldface center of the answer)—and the latter now appears ~~*struckthrough*~~. The third, middle ***who*** is our problem, as I will suggest imminently.

These new discoveries bring us to a crossroads: question expressions can travel quite a distance, but not in every context. We would like to formulate general rules that make every good question possible, no matter how many embeddings the question expression travels through. Yet, at the same time we want this rule to mark questions like 11Q* and 12Q* as bad, because this is how native speakers judge them. Importantly, the reason for the ungrammaticality of the latter seems to be syntactic, not semantic: it is easy to see that the ungrammatical monsters 11Q* and 12Q* do convey coherent meanings (as seen by their paraphrases 11Q✓ and 12Q✓). The system we built is not working, and it needs to be repaired. In sum, at least with respect to the puzzle in 11 and 12, the prime suspect for disrupting the Q* may now be identified: it is the second token of ***who*** that seems to be culpable. Once the arrow crosses it, the sentence is rendered ungrammatical, and both sentence and arrow earn a star.

Once again, these observations hit the core point. If any sequence of syllables constitutes a good word, we don't need a grammar; likewise, if a question can be asked by any sequence of words, a syntax is unnecessary. On the other hand, to rule out sequences of syllables and words such as 11Q* and 12Q*, which are deemed unacceptable by native speakers of English, rules must be invoked. Words and sentences are constrained as part of any speaker's knowledge, and it is our task to understand why this is so and to construct explanations for these phenomena. So, let's carry on.

From "Anything Goes" to Constrained Rules

Ungrammatical sequences are nothing new. Yet the discovery of *systematic* ungrammaticality the in the context of transformations, as presented above, was an important landmark in the development of the theory of syntax some 60 years ago.[3] It taught the field an important lesson. A theory in which anything goes, that is, where you can write rules at will as long as they do the job, comes with a heavy price tag: If you let transformations apply freely, your grammar produces monsters (this is called the "overgeneration" problem, as the grammar is too permissive and allows, or generates, too many possible structures). One voice may suggest a quest for ways to tame our rules, in order to block monsters like 11Q* and 12Q*. An opposing voice, however, might retort, "Why bother?" Why waste time on monster blocking, given that monsters hardly ever appear in texts or conversations? After all, at its daily toil, a language processing machine with a loose, unconstrained grammar never encounters monsters like 11Q* and 12Q*; so why are we dwelling on this marginal issue?

Here is my short answer, on which I will elaborate in part III: despite many similarities, language tool building and linguistic theorizing are not the same. The engineering effort is about machines that perform *actual* tasks, whereas science seeks the true explanation of human cognitive structure through inquiries into the full range of humanly *possible* performances. These two approaches may not always invoke the same considerations. In the present case, the contrasts in human judgments that I described—between legitimate long-distance travels of question expressions and a ban on such travels in certain contexts—reflect what humans know and can do (indeed, these examples come from grammatical testimonies of human owners of adult linguistic cognition). As such, they are at the center of scientific inquiry and demand an explanation, even though they may be rare, hence of little interest to machine builders.

Later I will reflect on whether my perspective on the commitments of a scientific linguistic theory is hopelessly romantic. For now, let's be more concrete: we have an urgent need to block 11Q* and 12Q* and deem them ungrammatical, while preserving all the grammatical questions. The tools we currently have at our disposal cannot do this. Indeed, the discovery of monsters like these led to an important shift in theoretical focus, which began in the late 1960s. Since then, a thorough investigation has attempted to uncover precise *constraints* on the format and mode of application of both syntactic transformations and phrase structure rules. These constraints block monsters, and they also allow rules to be stated more generally. The end result has been a new cognitive architecture for the grammar, which I will now motivate, explain, and then illustrate.

In discussing transformations and embedding (5–8) and attempting to formulate the rules (9 and 10), I considered a transformation that contained an unspecified element that I represented as ". . ." The idea was to allow the transformation to cross "any otherwise legitimate sequence of words." Yet this led to an overly permissive rule that admitted monsters (11Q* and

12Q*). The next move is to impose a constraint on transformations, one that would selectively bar them from applying in certain contexts: for example, if their application would lead to the crossing of a ***who*** or a ***what***. A way to implement this is to state the contexts in which transformations may apply and to bar them otherwise.

> In reflecting on the way theoretical constraints on transformations should be stated, several strategies present themselves. One way to formulate constraints on a syntactic system is to restrict a rule's domain of application; namely, to specify the conditions that must be met for a rule to apply (in its mapping from "deep" onto "surface" representations). An alternative strategy is to allow rules to apply freely but set limits on the format of the resulting representations: that is, to apply a kind of filter that looks at the output of rules and sifts out deficient results. Choosing between these possible methods involves serious considerations with far-reaching consequences. Over the years, syntacticians have provided precise answers.

The introduction of constraints on movement has led to generalization. Specific transformations (for questions, passive, etc.) were subsumed by the more abstract and generalized rule MOVE, an operation that copies elements from one position in a representation onto another and that has the capacity to delete the original copy (e.g., *What did John eat ~~what~~?*). The generalization ranges over different structures: for example, *That steak, John ate ~~that steak~~ yesterday*; *It is my soup that John ate ~~my soup~~ yesterday*; *What did John say that Bill heard that Mary is ~~what~~?* And much more, within a language and across languages.

By the same logic, phrase structure rules were subjected to a severe constraint (which will be revisited in chapter 5) that limited their format. This constraint later helped to formulate a single rule MERGE, which takes two distinct pieces in a representation and merges them into a single syntactic unit. It can do so to sublexical units, to single words, and also to phrases. By this theoretical development, many rules that could earlier be part of the human cognitive syntax were now blocked. The resulting theory, which is at the heart of an approach known as Minimalist Syntax, is much more constrained than its previous incarnations.[4] Other approaches to syntax may use different names and methods, but the idea is one and the same: constrain, simplify, and hence become more explanatory.[5]

How can a list of specific rules be squeezed into two? The answer lies in the architecture of the grammar: it now consists of the two rule types just discussed (though not formally presented), namely MERGE and MOVE, along with some other ingredients that will be laid out shortly. Critically, each sentence is associated with several highly constrained representations, mapped onto one another by these rules. The constraints set clear bounds on the notion "possible grammatical rule" and, in doing so, allow a restricted set of rules to be formulated. When stated correctly, they block monsters like 11Q* and 12Q*. Constraints on representations are key to the success of the grammatical system in assigning a correct structural description to any grammatically acceptable sentence and rejecting all ungrammatical strings.

An Example Constraint: Relativized Minimality

The ungrammatical questions (Q*) in 11 and 12 were said to be bad because the distance between the question word and its original position was too large. But how to define the notion "too large a distance" precisely? It must be done in such a way as to allow the acceptable travel of a question word out of multiple embeddings (5–8) but block the bad questions.

We noticed that in 11Q* and 12Q*, the link that the transformation formed between the question word ***what*** and its ~~*struckthrough*~~ original position crosses another ***who.*** An influential proposal by University of Siena's linguist Luigi Rizzi (here simplified) turned this observation into a general constraint on representations containing a transformational relation; this constraint, stated in 13, would derive the facts.[6]

13. **Relativized Minimality**

In the configuration . . . X . . . Z . . . Y . . .

Y cannot be related to X by MOVE if Z *intervenes* and Z is of the *same type* as X. Z will be annotated *IV* (for *intervenor*).

This formulation prohibits the presence of an intervenor between two MOVE-related positions, or else the resulting representation would be ungrammatical. Above, Z is present between X and Y. But to count as *intervenor* (*IV*), Z must be relativized to X, Y (i.e., X, Y, Z must all be of the same type, and that is a complicated matter). In 14–16, I illustrate how the *intervenor* concept works. In each instance, I added a paraphrase, $Q_{\checkmark}$, showing that the ungrammatical output of the transformation, or MOVE, nonetheless conveys a coherent meaning.

(continued)

14. A: I wonder who could solve the problem **in this manner.**

Q*: ****How*** do you wonder ***who*** could solve this problem ~~*how*~~?

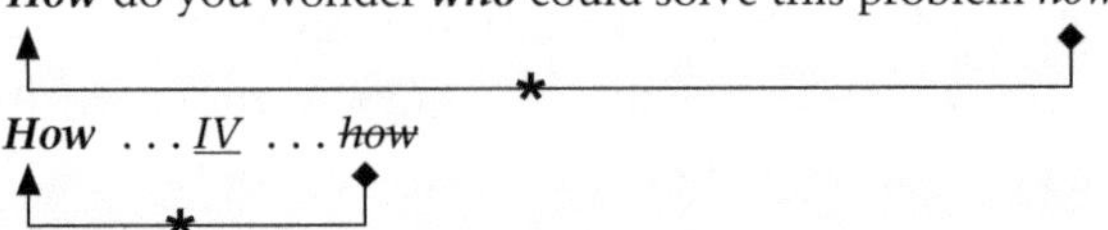

Q✓: ***What*** is the manner by which someone, about *whose* identity you were wondering, could solve this problem?

15. A: The dog which **Mary** saw was scary.

Q*: ****Who*** did the dog ***which*** ~~*who*~~ saw was scary?

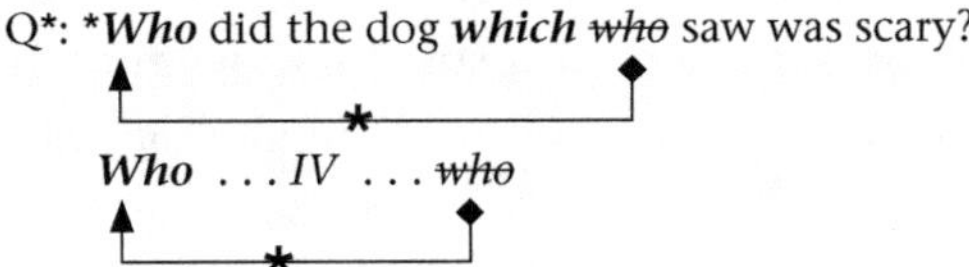

Q✓: ***Who*** is the person such that the dog *which* she saw was scary?

Relativized Minimality is violated by 14Q*, because ***how*** (= X) is related to ~~*how*~~ (= Y) by MOVE (a rule), where *who* (= Z) is an intervenor; X, Y, and Z are question words, hence of the same type. The string is thereby rendered ungrammatical. Similar considerations hold for 15.

Relativized Minimality covers a set of facts that seem unrelated and yet fall under the same generalization:

16. A: They could have left.

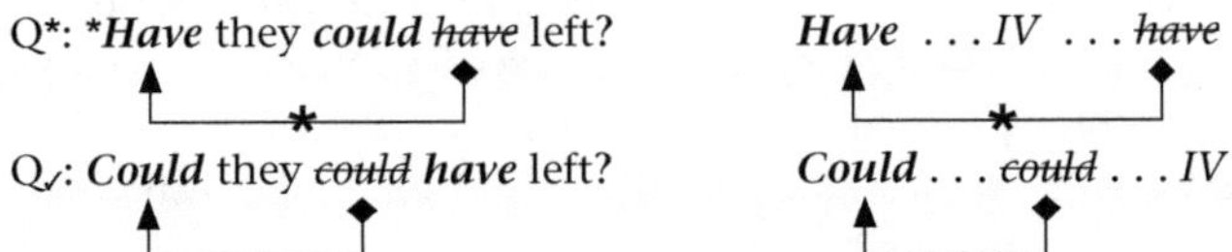

In 16Q*, X, Y = *have* and Z = *could*. The link between X and Y crosses the intervenor Z, and ungrammaticality follows. Yet in 16Q✓, X, Y = *could*, the relevant category Z = *have* is not an intervenor, and the result is a grammatical question. Surprisingly, then, Relativized Minimality covers such questions, too, and reveals itself as a constraint on representations with broad empirical coverage.

More Dependency Relations: *A Student Is . . .* or *A Student Are?*

The relation between positions in a sentence that is captured by MOVE turned out to be a central one in syntax, with many implications for the schema of the new cognitive syntax. It has also had ramifications for how humans process sentences in real time—ramifications that have been thoroughly investigated. But MOVE is not the only dependency relation one encounters in sentences. These relations are important, especially in the context of LLMs, so I will elaborate a bit. First among them is the familiar person–gender–number agreement between a subject and a verb in English (and between a verb and its object in languages like Italian). Person, gender, and number are taken to be features that typically cluster together (hence the reference to "person–gender–number" agreement), and they manifest in many of the world's languages. Even in a poorly inflected language like English, we say *This student is smart* but never **This student are smart* (here, I assume that if a student chooses *they* as their personal pronoun, then this sentence takes the form *These students are smart*). But what makes the starred one unacceptable? What rule or constraint does this sequence of words violate? Needed is a constraint that prohibits a lack of number congruence (**This student*$_{\text{Singular}}$ . . . + . . . *are*$_{\text{Plural}}$). While this appears to be a simple formulation, we must be clear about what ". . ." can possibly be. That is, can any material that intervenes between a subject and a verb in English substitute for ". . ."? The answer is in the negative. Here is a small demonstration. It appears that ". . ." cannot always stand for *nothing*. That is, mere linear proximity between a noun and a verb does not ensure that they must agree in person, gender, and number. This can be seen in the contrasts below (where agreement is annotated with dashed arrows, to underscore that the relevant dependency relation may be a relation other than MOVE).

17. **Subject–verb agreement (N$_{\text{Singular}}$ . . . N$_{\text{Plural}}$ . . . V$_{\text{Singular}}$)**
    a. ****A student***$_{\text{Singular}}$ who visited *us*$_{\text{Plural}}$ ***work***$_{\text{Plural}}$ very hard.
    b. ***Students***$_{\text{Plural}}$ who visited *John*$_{\text{Singular}}$ ***work***$_{\text{Plural}}$ very hard.

Why aren't the nouns closest to the verb agreeing with the verb in number? The answer lies in the hierarchy seen in figure 2.1. The embedded sentence, Sentence$_2$, is within the main sentence, Sentence$_1$, which describes the subject *student(s)*—it refers to those students who visited John and to no other student. The underlined verb should therefore agree with its (underlined) subject, which is linearly distant but structurally on a par. The adjacency between the noun *John* and the verb *work* is not a determinant in the subject–verb agreement rule.

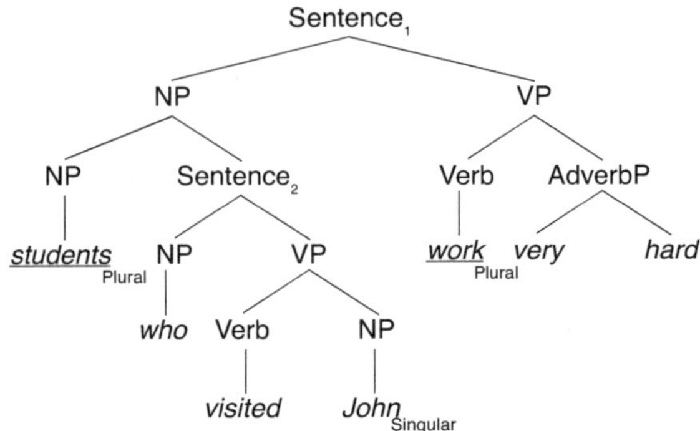

Figure 2.1
Embedded nouns may not agree in number with the closest verb. While the noun *John* is adjacent to the verb *work*, the latter agrees with the distant *students*, because they are on a par hierarchically (parts of immediate constituents of Sentence_1).

Do note that while the agreement relation is subtle, it may not be related to MOVE, which governs the relation between a displaced element and some other position. By what we have seen thus far, agreement is a relation between two stationary words or phrases from different categories, hence potentially independent of MOVE.

In sum, we have seen the complexity of the agreement dependency, its critical dependence on hierarchical structure, and its distinctness from MOVE. People occasionally make agreement errors when they produce such sentences, but for the most part, these errors are immediately recognized as such. The rule is clearly followed by native speakers, who can readily acknowledge the ungrammaticality of 17a. Upon careful examination, the agreement relation seems to not be subject to the Relativized Minimality constraint, which leads to the conclusion that it is not MOVE but constitutes, rather, a distinct dependency relation that requires separate regulation.

I know that keeping track of all the notation I introduced is not easy. But syntax—that piece of knowledge that our cognitive system possesses—is complex, and setting constraints on its rules is even harder. We shall soon see how this recognized complexity has led to design changes in the model for our cognitive syntax.

Even More: Did *She Look at Herself* or *at Her*?

A knight and three princesses (and no one else) are standing in front of a large mirror, and the knight says "Look at your own image, please!" Sentence 18a is an accurate report of what happened, indicating that ***herself*** picks up its reference from its most closely neighboring noun phrase ***every princess***. Curiously, this cannot be said of 18b, which may describe a scenario in which the knight's request was that every princess should look at some other unmentioned female (though there are none in the image), not at herself. Again, the dashed arrows stand for a dependency relation that is potentially different from MOVE.

18. a. The knight asked ***every princess*** to look at ***herself*** in the mirror.

    b. The knight asked ***every princess*** to look at ***her*** in the mirror.

Now imagine that a (distinct) maid is standing near every princess, and the knight is telling each maid to look at her princess's image in the mirror. The event cannot be described by 19a (as ***herself*** can only refer to *the maid*). However, 19b is an accurate report of what happened.

19. a. The knight asked ***the maid*** standing near *every princess* to look at ***herself*** in the mirror.

    b. The knight asked *the maid* standing near ***every princess*** to look at ***her*** in the mirror.

Now this is really peculiar: each pair of sentences conveys roughly the same message, but the way native English speakers interpret the dependency between ***herself/her*** and neighboring nouns turns out to be subtle: in the pair 18, ***herself*** must refer to its nearest noun, whereas ***her*** cannot (hence the starred arrow), but it can refer to a more distant one; in the pair 19, the pattern is different. Curiously, the linear distance between ***herself/her*** and ***every princess*** is identical in all four cases: they are separated by the three words *to look at*. The possible and impossible readings, marked by the arrows and stars, present a subtle, highly restricted pattern, which we cannot attribute to mere linear distance.

I will not burden you with a detailed syntactic explanation of this pattern (it can be found in any decent syntax primer) and will just say this: whatever the correct explanation of the peculiar distribution of pronouns and reflexives may end up being, it must rely on an analysis that assumes a structural hierarchy between parts of a sentence and highly constrained principles of shared reference between (reflexive and nonreflexive) pronouns like *her*, *herself*, and other nouns around them.

Interestingly, this set of restrictions on the referential dependency of pronouns interacts with other parts of syntax. This will become very important soon, when we come to the functions of "attention heads" in LLMs; they are entrusted with the task of linking positions within a sentence. Can they provide an account for the subtleties presented? We shall soon find out.

A Final Dependency: Ellipsis, at the Meeting Point Between Form and Meaning

Our last dependency relation displays fascinating interactions with the previous ones. At issue are incomplete sentences (a.k.a. ellipsis), in which the missing words—call this "a gap"—are not heard but nevertheless can be completed in the speaker's mind, as determined by a previous sentence. What is special here is that this completion is highly constrained—not just any completion will do:[7]

20. a. John devoured a cake, *but* Bill didn't devour a cake.
    (No ellipsis → Bill devoured no cake)
    b. John devoured a cake, *but* Bill didn't __.
    (Ellipsis → Bill devoured no cake)
    c. John devoured ice cream, *and* Bill __ cake.
    (Ellipsis → John and Bill devoured sweets)

Importantly, despite the missing pieces, no ambiguity is noted in any of these sentences. Although Bill's negated action in 20b is not mentioned, and yet the sentence *only* has the meaning implied in parentheses. It cannot mean that Bill did not bake a cake or that he did not devour a steak. The same meaning uniqueness holds of 20c—John and Bill ate different foodstuffs, but the reconstructed meaning can place no verb other than *devoured* in the blank, which is easily completed by the hearer. Whatever the process of meaning reconstruction in speakers' heads may be, its results in these cases appear to be strict. As with questions, we utter and hear such sentences daily and are able to complete the missing part easily and without error or hesitation. How can we formulate the meaning reconstruction algorithm correctly to ensure that the meaning of the missing piece is always unique? Representing it in a somewhat richer way, we replace the gap with a whole struck through Verb Phrase in 21b, (without its past tense marker, which remains on *didn't*), and a Verb in 21c:

21. a. John devoured a cake, *but* Bill didn't devour a cake.
    b. John devoured a cake, *but* Bill didn't ~~devour a cake~~.
    c. John devoured ice cream, *and* Bill ~~devoured~~ cake.

I can see a reader nodding and thinking: what's the fuss about? A simple rule that says *Complete whatever is missing according to the preceding clause!* would do. But wait: how exactly is a grammatical representation obtained? For 21b, the process must begin with the identification of the missing material in the second clause, which would somehow lead to the conclusion that the VP is missing; then, it would search for its parallel in the first clause ([$_{VP}$ *devoured a cake*]), strip it off its tense, add an auxiliary with that tense (*didn't*); finally, copy the resulting material (*devour a cake*) into the abstract representation of the whole sentence. For 21c, the completion would require the identification of the verb as missing in the second clause, its copying from the first clause ([$_{V}$ *devoured*]), and inserting it in the right place in the second clause.

This complex process is the minimum required for successful language use of such sentences, and it is a sequence of operations that speakers perform routinely and rapidly with novel ellipsis sentences. Moreover, writing a program that does all that reliably with any selection of words may not be easy at all, mostly because providing a general yet precise definition of the piece that must be copied into the ellipsis is difficult. The above description is incomplete, as well. To see that, consider the ellipsis sentence in 22. Unlike 20, this ellipsis contains an ambiguity: the pronoun in the ellipsis may connect to more than one antecedent. But most importantly, the ambiguity is highly constrained:

22. John walked his dog, *and* Mary and Bill did, too.

As any native speaker of English would agree, 22 may mean:

- Mary and Bill each walked his/her respective dog.
- Mary and Bill both walked John's dog.

But not just *any* antecedent for the pronoun will do—22 may *not* mean:

- Bill walked John's dog, and Mary walked Bill's dog.
- Bill walked John's dog, and Mary walked her own dog.
- Bill walked Mary's dog, and Mary walked her own dog.
- Bill walked Mary's dog, and Mary walked John's dog.
- Bill walked Mary's dog, and Mary walked Bill's dog.

Thus, on the one hand, ellipsis requires that the missing material would be filled by an exact copy that comes from the first clause; on the other hand, we see that sometimes, a constrained ambiguity is found. A theory of human linguistic ability must account for these highly systematic peculiarities, which we are all aware of automatically, as it were. I will stop short of providing an explanation, as it requires advanced syntax and semantics, but will only point out that, unlike simple cases of MERGE and MOVE, one must invoke interactions between the pieces in order to explain such odd phenomena.

The Cognitive Architecture of TGG—Take Two

Let's summarize this more technical chapter on linguistics. We have seen two generalized and highly constrained grammatical rules, MERGE and MOVE. The former builds hierarchical structure, and the latter establishes dependencies between positions in that hierarchical (tree) structure. We have also seen small hints of the need for additional operations, which may not involve displacement but do represent dependencies between positions in a tree (agreement, pronoun–antecedent dependencies, and the interactions seen in ellipsis). Reformulations of rules and constraints, with ever-increasing precision and generality, led over the years to predictions that could be tested, as further important linguistic phenomena were being uncovered, which motivated further theoretical modifications and extensions. In this way, two interfaces with syntax were established: one that makes contact with the logical and reasoning parts of our mental life, the other, with the physical qualities of speech. In some sense, this schema determines the flow of information from an idea to an utterance. Each of the interfaces is implemented as a linguistic level of representation, with rules and constraints of its own. To better understand how this computational machinery works, what constrains it, what empirical discoveries motivate it, and what accomplishments it can be credited with, we need to display the context of the computational medium within which the rules operate. The changes I just described and motivated lead to a new cognitive architecture, depicted in figure 2.2.

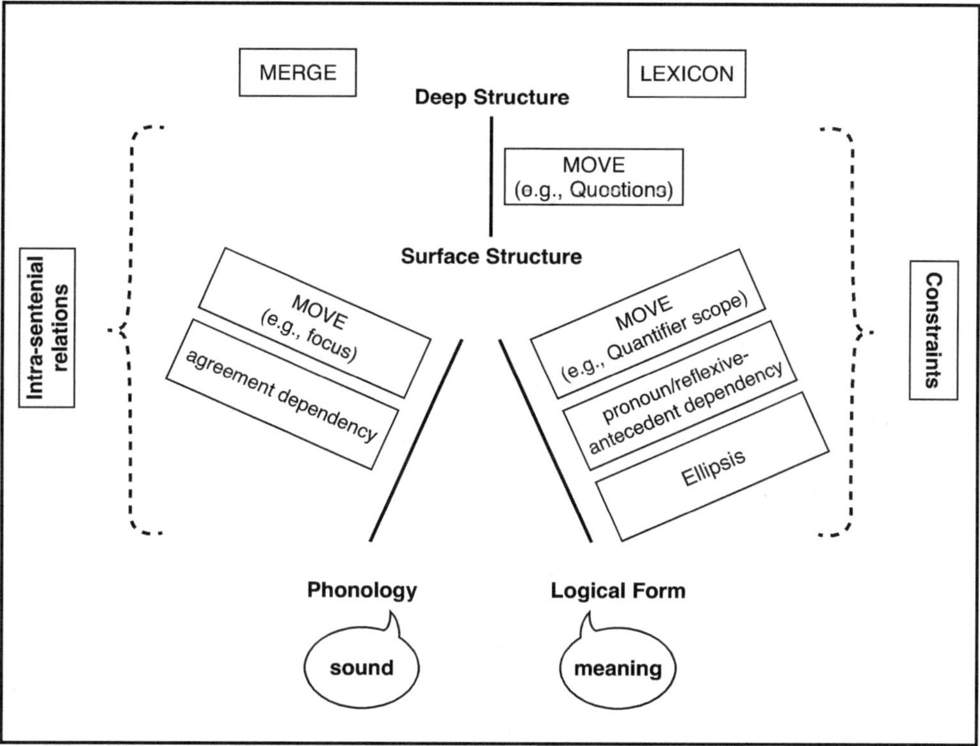

Figure 2.2
The Take Two model contains four levels of representation with which each sentence is said to be associated. Slanted boxes mark the rules that map from one level to another. MERGE applies to lexical entries drawn out of the lexicon. MOVE is an available mapping throughout; other rules are restricted to a particular mapping. The model is richer than its predecessor, yet it is much more constrained and has broader empirical coverage.

Figure 2.2 is a rich but, nonetheless, simplified and hence incomplete representation of the mental structure of syntax in its final, steady-state, adult form. Our mind and brain have representational media to accommodate four levels. Deep Structure is where a hierarchical representation is created when phrase structure rules (now formulated in a more abstract and generalized form as the rule MERGE) apply to the abstract representation of words, each projected from the mental lexicon. The output of this operation, a hierarchical "tree" representation with lexical entries as its "leaves," is passed on to the Surface Structure level. This process may or may not apply a transformational rule (MOVE) to the representation. From this level, the representation is mapped onto the meaning-related level of Logical Form (a term borrowed from Bertrand Russell) and, at the same time, to a sound-related form, Phonological Form. In each mapping, certain operations (in boxes) may apply. Critically, their application is highly constrained and theoretically motivated. Each sentence is associated with the four complex representations. But these are just names. In fact, other linguistic frameworks differ in their nomenclature, but as far as this book is concerned, there appears to be general agreement on the central ideas, including an architecture with levels of representation, recursive rule systems that build up hierarchies out of labeled elements, formal mapping rules, and constraints. There also seems to be a consensus on the nature of the explananda, namely the linguistic knowledge and behavior of native speakers. At the core of the discussion are therefore the tight constraints on each representation and each mapping (meaning that not just *any* representation or transformational mapping is possible) and the phenomena they are designed to explain. We will soon see how this architecture plays a role in psychological and neurological studies of language.

If you got this far, you should now have at least a bit of a flavor of what the theory of syntax is about, and why the study of its intricacies is important if we want to understand human language abilities. This conclusion holds even before other important topics are discussed: left untouched in this brief review are relations that are at the syntax–semantics interface; semantics (the study of how single meanings compose into more complex ones); and pragmatics (the study of that which is unsaid but implied). Some of these issues will come up in later chapters.

Taking Stock

In this chapter we saw some details about the architecture of our cognitive syntax:

- Abstract theoretical concepts that a speaker allegedly possesses: not only transformations but also constraints on transformations
- The move to a highly abstract theory with a minimal set of rules that are highly constrained
- An example: Relativized Minimality—a constraint on MOVE
- Other dependency relations: agreement and pronoun–antecedent relations
- Complex puzzles that ellipsis structures reveal
- The new cognitive architecture of syntax—Take Two—Deep and Surface Structure, Logical Form, Phonological Form, and the rules MERGE and MOVE and their varied instantiations—and some facts these elements are designed to explain

Part II Where We Are: Language Technologies

3 From Early Perceptrons to Deep Learning

The Brain—is wider than the Sky—
For—put them side by side—
The one the other will contain
With ease—and You—beside—
—Emily Dickinson, 1862

The essence of behaviorism is the belief that the study of man will reveal nothing except what is adequately describable in the concepts of mechanics and chemistry.
—Karl Lashley, "The Behavioristic Interpretation of Consciousness," 1928

Nongrammatical Alternatives

The plot that has unfolded so far goes as follows. We humans communicate linguistically by virtue of several rule types: morphophonology (resulting in plurals like *bus-ez* or *sock-s*, but not *bus-s* or *sock-ez*), syntax (*Could they have run?* but not **Have they could run?*), semantics (*It is not the case that John didn't sleep last night* means the same as *John slept last night*), and pragmatics, which I did not discuss (*John has five children* is taken to mean that he has *exactly* five children, although this is not explicitly articulated).

In the foregoing, we've focused on grammatical knowledge. Two critical questions were set aside. First, how does this rich knowledge base develop in the individual heads of all English speakers (is it innate, or is it somehow acquired, and if so, how)? Second, how do speakers recruit this knowledge when speaking or listening to others speak?

It is clear that for language development and the practice of language use, we are dependent on our brain. Our main question may therefore be: **Does our brain *really* possess and follow the rules of grammar**? TGG answered this question affirmatively, but it offered no implementation. The variety of machine learning approaches which I present below, and especially current day LLMs, answer this question with a flat **no.** The claims behind these models thus imply that the theory we have discussed is at best epiphenomenal, and they seem to offer the language sciences a fresh start. As will soon unfold, LLMs' aims, subject matter, and empirical evidence indeed differ in interesting ways from those of the linguistic approach. Linguistic research was presented as a scientific journey into the cognitive architecture of language; this chapter, by contrast, will be about language technologies and their applications, in keeping with this rapidly developing field. Are these perspectives really different? We'll soon see.

The story begins at a basic level, that of individual nerve cells and descriptions of their electrical activity. It then continues to neural networks that are capable of learning, and it travels all the way up, to the remarkable success that LLMs have had in the scientific exploration of the brain, especially the visual system. Can this success extend to language, as LLM proponents say? The answer is complicated, at the very least. But first things first; let's begin with the basic unit of a neural network, the neuron.

Origins of Computational Neuroscience—"The Neuron Doctrine"

Neurons are universally recognized as distinct, naturally occurring biological entities. This conception has been crucial to the development of the neural networks project, and hence it is a good place to begin. But the fact that neurons are distinct cellular entities, massively connected to others in special ways, was officially recognized only in 1906. Two eminent anatomists, Santiago Ramón y Cajal and Camillo Golgi, were awarded the Nobel Prize in Physiology or Medicine for that year, in appreciation of methods and discoveries relating to the brain's neural network.[1] From that time on, the scientific community has supported the notion that the brain is made out of cells, with neurons playing a central role in its functioning. It may be odd to think that as late as the beginning of the twentieth century, the idea of the neuron was a mere "doctrine"—a belief you may hold about the real world, but one lacking empirical justification. Indeed, convincing the scientific community that the brain is built out of cells was difficult, because light microscopy failed to provide a good view of brain tissue. It thus took Golgi's and Ramón y Cajal's imagination, technique and tenacity to convince the world that like other bodily organs, the brain is made out of cells.

Once the neuron was recognized as a basic unit, scientists began studying its properties, initially by measuring its electrical activity. They understood that the effect of neurons on behavior (from vision to language behavior, abstract and complex as it may be) is what their theory seeks to explain. But how do neurons work, and how do we model them? Contributions have come from a variety of fields: mathematics, neurophysiology, electrical engineering, linguistics, and psychology. These have colluded to build a discipline that has become known as computational neuroscience.

I begin this chapter with the story of the first, twentieth-century attempts to theorize about neurons. I then move on to network models of neuronal aggregates, which have focused on the visual system. Finally, I discuss past and current deep neural networks. This will serve as an introduction to computational approaches to language, including LLMs, which will conclude this short journey. As before, precision and theoretical detail fall victim to the need for clarity, as the challenge is to present LLMs and their precursor systems at a level of detail that is sufficient for an intuitive understanding of their overall structure, using no mathematical notation. Hopefully, a reader will get enough of an idea of the main tenets of LLMs to make subsequent discussion comprehensible and effective. Here, too, slightly more technical issues are boxed, and while readers are encouraged to brave these, flow-seeking ones may skip them, hopefully without loss of continuity.

Early Neural Network Models—The McCulloch–Pitts Neuron and Perceptrons

It took forty years from the time Ramón y Cajal and Golgi's work on the neuron was recognized to the first serious mathematical characterization of this biological entity.[2] Physician Warren McCulloch and logician Walter Pitts presented a model that drew a connection between logical circuits and networks of neurons. The phenomenon of all-or-none neuronal spiking was already recognized, and on its basis, McCulloch and Pitts designed a network of small computational devices that resembled neurons in being binary and interconnected and in having a fixed spiking threshold (the voltage on a neuron's membrane that's required for firing); they showed that this network could solve logic problems.

Each neuron in the McCulloch–Pitts model has two states (0, 1) and is fed by many others with excitatory and inhibitory inputs. All inputs have equal weights. Excitatory inputs are aggregated in the neuron, which moves to state 1 and fires if its excitatory level exceeds a given threshold θ. If the aggregated excitatory input is below θ or if it contains *any* inhibitory input, the neuron remains silent at state 0.

McCulloch and Pitts claimed that through a choice of different θ values, this simple computational device could calculate any Boolean function (i.e., a function that operates on logical values with binary variables). That is, given n inputs, it could calculate NOT (with $\theta=0$ and inhibitory input), AND, and OR (with excitatory input and $\theta=\text{n}$ and $\theta=1$, respectively); it could thus calculate NOT(OR) and especially NOT(AND) = NAND, a key component in basic logic circuits, hence familiar to engineers.

The McCulloch–Pitts model was an important theoretical accomplishment, as it connected neurophysiology with logic and engineering for the first time. It demonstrated that a network of computational nodes that shares traits with the brain's basic unit can work together and model a logic circuit, which can aggregate input and produce a binary output (0, 1). Still, while being a major advancement, this "neuron" was nonetheless static. The weights on the connections were all equal and fixed, and the spiking threshold θ (known as the "bias" term) was set in advance. As a result, the mode of action of the "neuron" was unchangeable on input. In other words, the McCulloch–Pitts neuron could compute basic logical equivalences, but it lacked the capacity to learn.

Fifteen years later, psychologist Frank Rosenblatt wanted to build a machine that could be trained to identify images. The McCulloch–Pitts neuron served as his point of departure, and he proposed important modifications: he allowed the fixed thresholds and weights of input units to be varied and modifiable, and he canceled the absolute power of inhibitory "synapses."[3]

These new features provided flexibility, which opened the door to an algorithm that, given certain external inputs ("a training set" or "teacher"), could modify the values of the weights of incoming inputs. Rosenblatt understood that the modifiability of these weights would bring about a change in the machine's capabilities, and he put emphasis on this feature. In other words, he proposed a sketch of the first learning rule that was connected to a binary classifier—a device that could produce a yes/no (or true/false) decision in an image recognition task. He called the machine that materialized his ideas a *perceptron*, and he tried to implement a learning rule within it. As the purpose of the machine was to recognize images, he embedded the algorithm in a vision device. Using multiple detectors, he expected his perceptron to learn to identify objects in its visual field. He reasoned that a working perceptron, which would be built out of neuron-like units, would be a good model of the behaving brain, as it would illustrate how complex human behavior follows directly from the electrical activity of neuronal aggregates.

The ingredients of the first perceptron (without the learning rule) are shown in figure 3.1. The inputs from other units ("neurons") are weighted, summed, and added to the bias term θ, which can be chosen anew for each problem. This is converted into an output (0, 1).

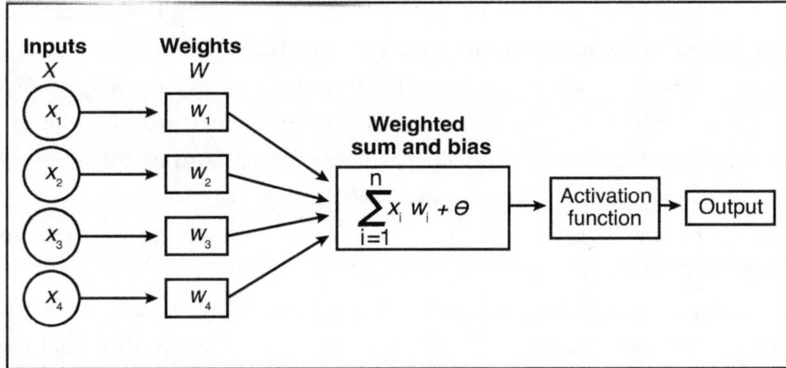

Figure 3.1
This perceptron has input and output layers, and a weight is associated with each input node; the input nodes feed into the unit, where all inputs are summed up and added to the bias term θ. This sum is passed through an activation function—a binary classifier whose job is to map a continuous numerical value onto 0 or 1 (true/false).

This setup has multiple parameters: the input column X has as many numbers as there are "neurons"; then there is a column of weights, which may now be variable, described by the vector W—a shorthand for the column of numbers it stands for, w_1, w_3, w_3, w_4, which is as long as the number of input units—and θ. The activation function (typically a sigmoid) determines the output, based on the weighted sum plus bias. (Again, no learning rule is represented in the diagram.)

Learning in a Perceptron?

A learning rule is valuable only if it is provably capable of learning its task on a reasonably small input set and within a reasonably short training cycle. But let's reflect on what learning in a perceptron might be. The schema in figure 3.1 shows inputs x_i being assigned different weights w_i as they are passed on to be summed up. Suppose that we set all weights at the same value (let's say that $w_1 = w_2 = w_3 = w_4 = 1$); this means that the contributions of all inputs are equal. If this setup leads to a correct output, then the program works well, and no change in its internal state is necessary. But if the output on its first try is incorrect and if the program has a way of knowing it, then the way for it to try to improve its next output is to change the relative contributions of each input node, by changing their corresponding weights—perhaps x_1 is less important than the others, and thus its weight w_1 needs to be reduced? To illustrate, let's take the English plural example from chapter 1, and let's suppose that the program's task is to pluralize correctly two types of nouns: *drop*→*drop-s*; *clog*→*clog-z*. The program receives four letters for each noun (*drop*: $x_1 = d$, $x_2 = r$, $x_3 = o$, $x_4 = p$; *clog*: $x_1 = c$, $x_2 = l$, $x_3 = o$, $x_4 = g$, etc.), and in the absence of a rule (as the program has no access to abstract constructs such as "voiced consonant" and its effect on the plural *-s*), it must decide whether an *-s* or a *-z* needs to be added to the right of the sequence to produce a well-formed plural noun. Suppose further that we have a way of telling the program whether its current output is correct or incorrect and what to do in order to improve its performance on its next trial. The only modifiable elements are the weights on the connections of the network. If it starts off with the same weight assigned to all input letters, a correct answer in every trial is not more likely than pure guessing. Yet we need it to improve quickly. The English plural rule hangs on the noun's last letter, and it is on this letter that more weight should be put. Thus, weights w_1, w_2, w_3 should be minimized, because their inputs x_1, x_2, x_3 are not relevant to the phonetic shape of the plural; w_4 needs to be maximized, because it determines the correct choice. Weight modification would mean a change in the internal state of the program (as the weights would no longer be equal), which would improve its performance.

This move helps the program to focus on the most relevant element in the input sequence, but our problem is not solved yet: a relation between the input word's last consonant and the right choice between *-s* and *-z* needs to be established. Forcing a program, a machine, or an organism to change its internal state based on input amounts to *training* it; it learns, which is what we sought to accomplish in the first place. Rosenblatt understood this, and he claimed that once the right parameters were chosen, his algorithm converged on the correct weights, which meant that his perceptron could learn a rule on finite input and on a finite training cycle. The implication was that a computing machine built out of "neurons" can learn to perform a task by trial and error, presumably simulating a human thought process.

In keeping with his predecessors, Rosenblatt claimed that his algorithm had the capacity to implement any Boolean function (a function that can only return 0, 1). As it turned out, he was wrong. As a technology, performance turned out to be below expectations, and funding was soon terminated. Other ambitious, high visibility attempts to construct a rule-based machine translation device also resulted in failure, which led to their demise as well.[4] Not much was happening in machine learning in the 1960s.

Hidden Layers

About a decade later, AI pioneers Marvin Minsky and Seymour Papert of MIT turned to a limitation of the perceptron, the issue of *linear separability*.[5]

A function is linearly separable if, in a geometric representation of its outputs, a straight line can separate those assigned 1 and those assigned 0. The graphs in figure 3.2 below illustrate this with AND, OR, and XOR (exclusive OR—x OR y but not both). The figure displays two-dimensional representations with two parameters, x_1, x_2, that get values 0, 1 (an empty or full dot, respectively); x_1 AND x_2 is assigned 1 just in case $x_1=1$, $x_2=1$; x_1 OR x_2 is assigned 1 just in case $x_1=0$, $x_2=1$ or $x_1=1$, $x_2=0$ or $x_1=1$, $x_2=1$; finally, x_1 XOR x_2 is assigned 1 more narrowly, namely just in case $x_1=0$, $x_2=1$ or $x_1=1$, $x_2=0$. A diagonal line is the geometric representation of the linear classifier that separates AND and OR

(continued)

successfully (left and middle panels). However, no straight line can do that with XOR (right panel).

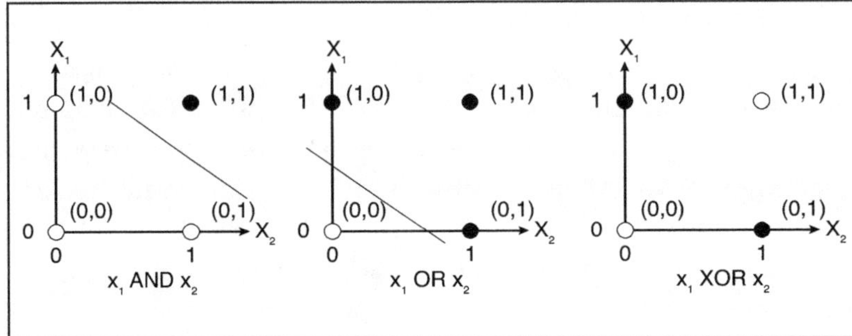

Figure 3.2
Linear separability of the logical functions AND, OR, and the inseparability of XOR: a straight line can separate the true formulas (full dots) from the false formulas (empty dots) for AND and OR but not for XOR.

Returning to the earlier goal of building a machine that calculates logical functions, Minsky and Papert demonstrated that Rosenblatt's perceptron, which had to decide whether an input is true or false (returning 1 or 0, respectively), could indeed implement Boolean functions such as AND and OR (inclusive OR, i.e., *x* or *y* or both) but failed with the function XOR (exclusive OR—*x* or *y* but not both), which is not linearly separable.

As is often the case, the discovery of an important limitation harbors rather than hinders progress. Minsky and Papert's proposed solution to the XOR problem was to extend the class of perceptrons beyond linearity. They showed that XOR is classifiable if a layer is added to the network; and not just any layer but one consisting of units that are neither input nor output layers. They proposed an intermediate and invisible, hence "hidden" layer. Such a layer has no access to the actual input or output; rather, it is completely locked in the machine, receiving a weighted sum of the input layer's output, which it passes on to the output layer.

This perceptron was indeed capable of a richer range of computations, but at a cost of increased network complexity: even if the number of units was kept unchanged, their interconnectivity grew by necessity, which is exactly what made the system go beyond linearity. Thus, Minsky and Papert's design laid the foundation for many future generations of neural network models. Still, the added complexity left many (Minsky and Papert included) in doubt about the network's theoretical validity. The addition of hidden units was moreover anathema to many associationist psychologists, who abhorred theoretical moves for which the evidence appeared to be indirect (an objection similar to the one that led them to dismiss Chomsky's TGG). To them, the scientific justification for a hidden layer relied on reasoning that was too complex and nontransparent. The network approach to the explanation of human behavior was left by the wayside, as even its progenitors preferred the symbolic path. The first round of attempts to build intelligent machines that learn was facing a dead end.

But there were those who nonetheless continued to investigate networks and their properties. The freedom to add hidden layers and choose different values for the weights of units opened the door to richer and more powerful designs. These were mostly based on modifications of network architecture—the number and relative size of the layers, their degree of interconnectivity, and so on—and subsequent tests of their performance. An illustration of a simple network with a hidden layer is in figure 3.3A; it features an input layer, a hidden layer, an output unit with an activation function, and full connectivity, as each node is connected to all the nodes of the subsequent layer (unlike Rosenblatt's network in figure 3.1, in which each node in the input layer is connected to a single node in the output layer).

Each unit $x_{i,\,j}$ has two indices: the first index, i, marks the layer that the unit is in (1 = input layer, 2 = hidden layer), and the second, j, marks its row within its layer (1 through 4). That is, $x_{i,\,j}$ is the unit in the i-th column and the j-th row. The values of the units in layer 1 are thus described by the vector $X_1 = (x_{1,\,1}, x_{1,\,2}, x_{1,\,3}, x_{1,\,4})$, whereas those in the hidden layer are described by $X_2 = (x_{2,\,1}, x_{2,\,2}, x_{2,\,3}, x_{2,\,4})$.

Now, to the weights on the connections. The first two layers of nodes in this network, the input layer and the hidden layer, are fully connected. Each connection has its weight, marked by $w_{i,\,j}$, where i = the position of the origin and j = the position of the destination. The arrangement of both panels of figure 3.3 displays such connectivity: in 3.3A, each unit in the input layer is connected to every unit in the hidden layer; in 3.3B—a network with an additional hidden layer, labeled "output layer"—full connectivity between the input layer and the hidden layer is depicted as well as between the latter and the output layer. The final connections (vector U in 3.3A, V in 3.3B) are partial.

(continued)

As weights are now allowed to vary. Listing W and U requires a somewhat cumbersome notation. Below, I illustrate with all the weights in W (this level of details is suppressed in the figure to avoid clutter):

$w_{1,\bullet}$: the weight on the connection between the first unit in the input layer $x_{1,1}$ and each unit in the hidden layer, $x_{2,1}$–$x_{2,4}$ $w_{1,1}$ $w_{1,2}$ $w_{1,3}$ $w_{1,4}$

$w_{2,\bullet}$: same for $x_{1,2}$, the second input unit, and the hidden $x_{2,1}$–$x_{2,4}$ $w_{2,1}$ $w_{2,2}$ $w_{2,3}$ $w_{2,4}$

$w_{3,\bullet}$: same for $x_{1,3}$, the third input unit, and the hidden $x_{2,1}$–$x_{2,4}$ $w_{3,1}$ $w_{3,2}$ $w_{3,3}$ $w_{3,4}$

$w_{4,\bullet}$: same for $x_{1,4}$, the fourth input unit, and the hidden $x_{2,1}$–$x_{2,4}$ $w_{4,1}$ $w_{4,2}$ $w_{4,3}$ $w_{4,4}$

If you took the pedantic route and followed the details, you noted that W is not a vector but rather a 4×4 matrix of parameters $w_{i,j}$, where $1 \leqslant i, j \leqslant 4$. We can think about each column or row as an ordered series of values or, better yet, as a vector, which together with the weights of the other units in the input layer forms a matrix W. Likewise for U in figure 3.3B. However, U in 3.3A and V in 3.3B are vectors, each consisting of four weights. The additional hidden layer in the network in 3.3B enhances its performance.

In massively connected networks, which are currently used in AI and which are also found in neural tissue, the weights on the connections, called "synapses," are crucial, and their values are often represented in matrices. This is where mathematicians and physicists make a contribution. Matrices are mathematical objects whose properties are well-understood. Neural networks can thus be analyzed, and conclusions and predictions that follow from these analyses can be pitted against behavioral and physiological experimental results, inspiring new experimental explorations. This has indeed happened since the mid-1980s; multilayered networks (as in the relatively simple one in figure 3.4) have been constructed and their classificatory abilities explored.

Note that we have thus far set no constraints on network connectivity. That is, no upper bound has been set on the number of units per layer, the number of layers, their similarity to one another, or the relation between rows and columns. All we know at this point (admittedly without much detail) is that, when input X passes through hidden layers, each with its own weights, the end result at the output layer is (likely) different from the input; and that this input–output difference is a result of the path the input

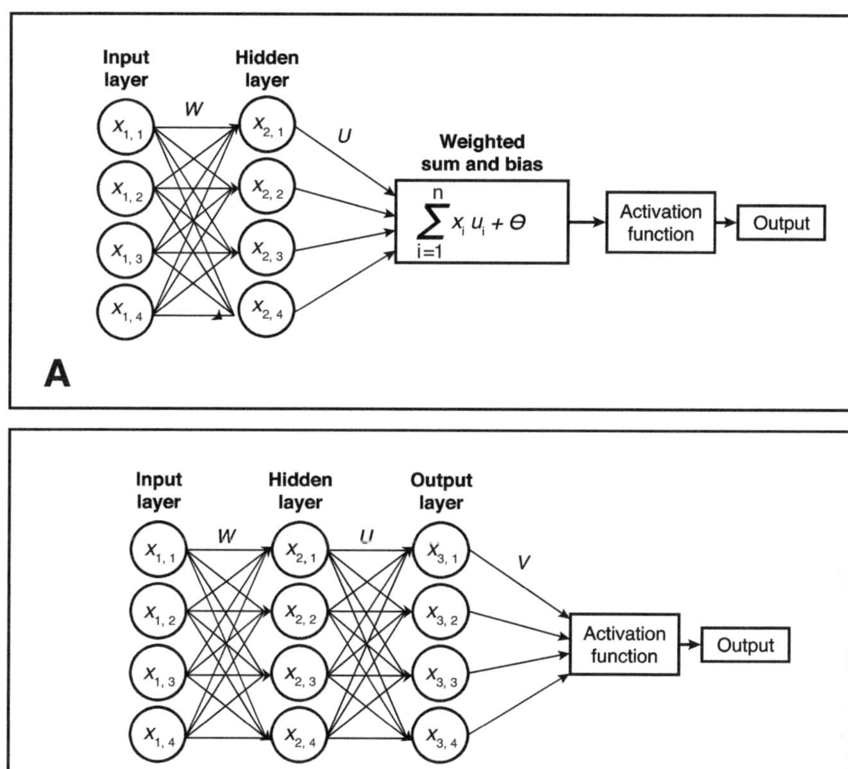

Figure 3.3
Perceptrons with one (A) and two (B) hidden layers.

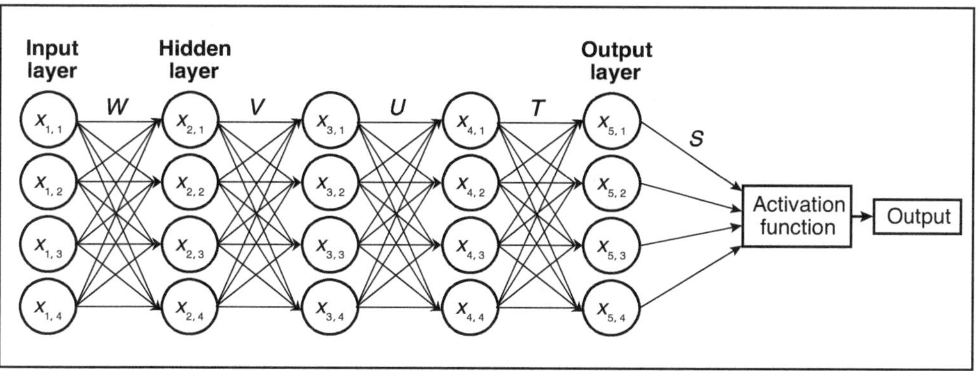

Figure 3.4
A multilayered network with weight matrices W, V, U, and T and vector S.

traverses as it is passed through the network's layers and on to the output terminal. This difference is a result of learning, which is, very generally, a change in a machine's state that is due to input. At this point, we don't know much about this input–output difference—how small or big it may be or how exactly it is brought about; but we have to acknowledge that given an input, any change in the machine's current state relative to its previous state would be indicative of learning.

Feedforward and Recurrent Neural Networks

In the above architectural designs, the unidirectionality of information flow stands out. All panels in figures 3.1–3.4 assume that the input data vector X is passed only forward (input units → hidden units → output unit). Such architectures are therefore called *feedforward networks*, and they seem to classify objects in static images at a high level of accuracy. Note that the graph in figure 3.4 allows full connectivity. That is, every node in layer 1 is connected to every node in layer 2 and so on. This may not always be the case. As we shall see later, networks that are not fully connected (a.k.a. *convolutional networks*) are closer to the biological reality of the visual system and moreover are constrained in terms of what they can learn, which is a virtue, as their power is not unlimited.

Feedforward networks don't work as well with nonstatic materials that unfold over time—tone by tone, frame by frame, or word by word—that is, with tasks in which the input's temporal structure matters. Such tasks harbor multiple connections between positions in the sequence, for which a local memory is critically needed. Engineering ingenuity augmented feedforward networks with additional components. A loop that connects a node to itself, and is fed by the value of its immediately preceding state, which may make a contribution to the node's current state, is added. Because the process recurs, these networks have been called *recurrent neural networks* (RNNs). An example of this class of networks is given in figure 3.5, in which every node in the Output layer is connected to itself (thick looped connections whose weights constitute the vector R=r_1, r_2, r_3, r_4). When we compute the value of any node at this layer (e.g., $x_{3,j}$) at time t, we take into account not only all the connections that pass information to this node now ($x_{2,1}$, $x_{2,2}$, $x_{2,3}$, $x_{2,4}$), but also, its own previous state at t –1, weighted by the values of R. Any node in a layer that has recurrence receives at time t both the

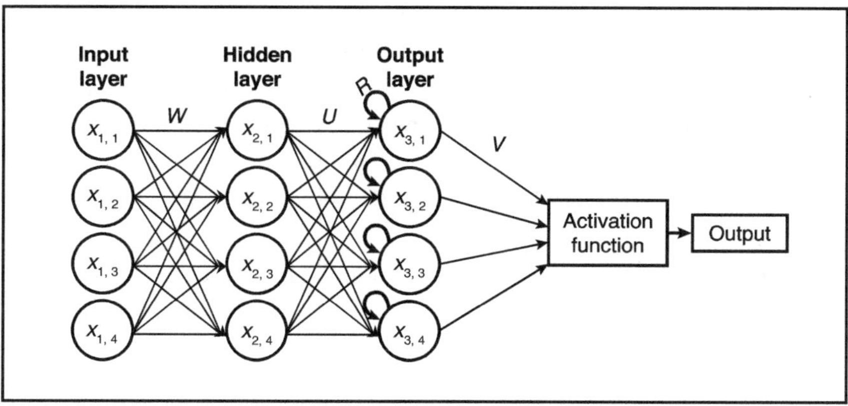

Figure 3.5
An RNN with weight vector *R* on the recurrent loop at the Output layer.

weighted input from previous layers and the weighted input from its own state $x_{i,j}$ at time $t - 1$. This recurrence serves as a memory of things past (at $t - 1$), which contributes to its present state *t*. Note that figure 3.5 features partial recurrence: the nodes that are fed not only by all others but also by themselves are only those in the Output layer.

Recurrence means that the network is no longer a feedforward network, but rather, one in which the flow of information is not unidirectional. RNNs thus help us to create a multiplicity of local memories at any given point during the processing of a sequence: one memory for a node's most recent state, one representing the relative contribution of both recent and less recent states, and so on, limited only by the machine's capacity to represent distant past state, known as the "vanishing gradient problem"—the further away you get from an input, the lesser its impact. A solution of this problem will need to weigh the importance and relevance of the current input against the importance and relevance of the context within which it is couched—the input that is more distant.

Our next step will be to extend the capacity of the memories RNNs avail us, which would optimally give a better expression of the possible of "past context." This will be captured by cleverly engineered mechanisms such as *attention* and the intentionally oxymoronic Long Short Term Memory (LSTM). Such mechanisms enhance the performance of these machines quite significantly, as will be shown below.

It is time to pause and see how far we've gotten. We have seen how the first quantitative model of a single neuron was enriched by the somewhat more complex architecture of the first perceptron, which functioned as a reasonable, but limited, classifier with minimal capacity to learn. Next came perceptrons with yet another architectural augmentation—a hidden layer—which further complicated the network but was more powerful, having gone beyond linearity. This move was historic, as it opened the way to massive connectivity, to enriched varieties of synaptic weights, which resulted in more powerful machines that could solve more problems. The feedforward nature of these networks was later augmented by loops serving as memories, which enhance their performance significantly, especially with input data that unfolds over time. These are RNNs.

This is also the time to wonder about the notion of Deep Neural Network (DNNs). What makes a neural network deep? In principle, any network with at least one intermediate, or hidden, layer (which is neither an input nor an output layer) is a deep network. But this name also implies architectural flexibility that enables the extraction of abstract, perhaps deeper, generalizations from the input. Depth, moreover, is a feature that might be closer to the actual biology; modeling brain processes may become more realistic when networks are allowed to be deeper.

Brain modeling will later be at the very center. For the moment, we stay at the machine level and move the time arrow ahead, which brings us to a critical issue: learning in networks. We have thus far assumed that the network is already trained, that is, that it has reached a steady state, hence the values of its weights are fixed. But little was said about *how* these parameters were set.

The Backpropagation Algorithm

In the animal kingdom, organisms learn through experience, whether supervised by an "instructor" (who informs them whether or not their output is correct) or unsupervised—by mere exposure to sufficient amounts of naturally occurring inputs. In machines, learning is described as a change in the machine's internal state, induced by input. In networks, such a change is brought about gradually through iterated trial and error. Typically, the network's initial output is compared to the expected behavior; if the match between the two is not perfect, an iterative process that adjusts the network's weight system, or parameters, is initiated. If the method that determines the adjusted weights is effective, then the level of error measured in the next round will be reduced. The process is then reiterated, until the error becomes minimal, and the adjustment of parameters is stopped. If this repeated modification of the network's internal state is successful, it will eventually perform as expected. Interestingly, this construal of learning by machines is in keeping with the idea that in the biological world, learning occurs through gradual adjustment of the strength of synaptic connections.[6] For learning machines, the readjustment of weights may be a possible solution for the learning problem that Frank Rosenblatt dreamed about. He called this process "backpropagation," but he was unable to formulate it.

How can the weights be modified to produce a network's correct performance? It took a long time before a solution was found. It was based on optimization techniques and clever mathematical tricks, assembled and connected to behavioral tasks. A landmark paper about backpropagation (or backprop, for short), which is most closely associated with this solution, was published in 1986.[7] It describes an algorithm that compares the actual performance of a layered network to its desired output, calculates the network's output error, and on this basis recomputes the weights for each layer. The paper then demonstrated that this algorithm successfully performs several tasks.

This demonstration was powerful and reignited interest in neural networks. Over 30 years later, one of the authors (Geoffrey Hinton), together with the two other progenitors of this approach and ideas (Yann LeCun and Yoshua Bengio), would win the prestigious Turing Award, for their contributions to machine learning.[8] Hinton would go on to win the 2024 Nobel Prize in Physics, though for a different innovation, as the Nobel citation notes (see prologue).

Let's examine backpropagation in some detail. Learning, as we said, is a change in the internal state of the network, as manifested by the values of its parameters—the weights on its connections. The goal is to teach, or "train," the network to perform exactly as expected, namely, until its output contains few, if any, errors. Error level is evaluated by comparison to a preexisting "gold" norm (the programmer's expectation of the network's ultimate performance). A clever algorithm determines small changes in the values of the network's weights, proceeding on a layer-by-layer basis, starting from the output layer and moving backwards. Once this readjustment process is completed, the network receives input again, its performance is once more evaluated for error against the golden norm, and the algorithm makes a slight readjustment of parameter values, bringing about yet another small change in its performance. This training process is reiterated up to a point where error level (the difference between the desired and the actual performance) can no longer be minimized. If error level approaches zero, we can say that the network has reached a steady state, having learned its task.

The introduction of the backprop algorithm moved the world of networks to a new phase, for this algorithm is no longer strictly feedforward. It allows, in fact requires, that at least during the training phase, signals propagate backwards in a manner that allows the weight values to be readjusted. This means that information flow is bidirectional.

Backprop begins with the evaluation of the network's output error (calculated by the distance between an input and the correct or expected output), which it proceeds to minimize. On this basis, it calculates and implements adjustments to the network's weights that are needed for error minimization. The algorithm iterates evaluation, adjustment, and retraining on more input, and it stops only when an optimal solution—output that is as close as possible to the target output—is reached.

How is this done? Prior to training, the network is in its initial state, where all weights are randomly set. The network's output vector is therefore incorrect (which would lead to mistaken classification by the activation function). The extent of the error is calculated by a similarity, or "loss," function that extracts the output vector (values of the units in the network's last layer) and compares it to the target, or "gold," vector that the designer provides. At this stage, the weights are fixed. The comparison gives a measure of similarity (or lack thereof) between the network's current performance and its goals.

The goal here is error minimization. Any chosen loss function has a minimum (at least one). Once loss is calculated, a path to reach the function's minimum needs to be drawn. In its search for the minimum, backprop is typically guided by an analytic method named gradient descent, that determines the fastest route to the minimum of the loss function. The search consists of evaluating the current weights in each layer and deriving adjusted weight values that will get the network closer to the desired minimum. This derivation and subsequent assignment are done backwards—from last layer to first. The readjustment of the weight values in all the hidden, intermediate layers is the hardest part: the algorithm must find new weight values for all hidden layers, and these depend on one another. The calculation process is complex, and proceeds from the Output layer backwards. A Chain Rule, invented for infinitesimal calculus by Leibnitz, enables the calculation of the rate of change of a function as a composition of change rates of a series of functions, which allows the breakdown of the calculation of weight values into a layer-by-layer process.

Backprop operates in cycles—training, error evaluation, and weight adjustment, followed by repeated training, evaluation and adjustment, and so on, until the loss function is minimized and the iterative process stops. Importantly, the effectiveness of the loss function depends on its having a single minimum. This is so because, with multiple minima, the system may get stuck in a local minimum that is not the desired one. Special measures are required to get around this problem.

The process—the loss function, the computation of gradient descent, and the calculation and subsequent readjustment of weights, followed by another input cycle—is reiterated until the minimum of the loss function is reached, and processing is then terminated. Backprop enabled the system to find a way to produce the correct results, as the network optimized its overall internal state.

Backprop applies to a rich variety of network architectures, and it exhibits an impressively broad range of abilities: from the identification of handwritten language to remarkable visual accomplishments. It was, without a doubt, a foundational step that brought us closer to the current LLMs.

Backprop and Questions About Learning

The central role of backprop in machine learning brings four important issues to the fore, especially in the context of natural language:

- *The nature of learning*. Backprop presupposes access to the target "gold" output. Error cannot be determined without such access, because a comparison between the network's output and the expected output is required. This comparison is known as "supervised learning," a process of change of the machine's internal state when the target is known. Learning may not always take place in this fashion, as has been repeatedly noted. Some learning is "unsupervised," occurring without correction: it is fairly clear that most, in fact almost all, syntactic development takes place in the absence of correction.[9]

- *Initial state*. As described, the weights in the initial, pretraining "knowledge state" of a network are randomly set. Does this feature correspond to an actual cognitive state? In the context of language, it might translate to a position on the Innateness Hypothesis, that is, on the question of whether humans possess knowledge about language that is not acquired through input.[10]
- *The manner of learning*. A commonsense, perhaps school-based, conception of learning (and especially teaching) is generalization. We conceive of learning by internalizing a rule as being more effective than learning by example. Machine learning experts are quick to point to the failure of past attempts to create AI-based technologies with rules, especially in the language domain. This is true, and what remains to be discussed is the significance of these failures for scientific explorations of language and linguistic ability.
- *The neurocognitive basis of backprop*. DNNs pride themselves of having much neurophysiological evidence that supports them (in chapter 8, I will review evidence for feedforward networks in the monkey's visual system). But as backprop requires bidirectional information flow, are neural systems in the biological world bidirectional? And when it comes to humans, we should ask: in what way, if any, is backprop related to the way we learn and behave? The answer to this question is far from clear.[11] An anecdotal but not unimportant hint regarding this question is Geoffrey Hinton's recent remark, "It's not clear that the brain can do backpropagation."[12] Here, too, the significance of these doubts about machines for scientific exploration needs to be discussed. We will do so soon.

2012 and the Explosion of Deep Learning

Superfast and more powerful computers now enable larger numbers of network nodes and layers. Larger storage capacities allow the size of the training sets to grow as well. Creative engineers have been trying out many network designs, with new mathematical and architectural tricks and innovations, which have been predicted to bring about computational explosions (some have actually occurred). The field has been brewing with more and more possibilities. But before we continue, one neglected but important point must be acknowledged: little was said above about *im*possibility—about what networks can*not* do. This was negligent, but for a reason: in 2012, everything about network performance changed, so there is no point in presenting the obsolete. The publication of AlexNet, an image classification network, hugely scaled up networks' performances as well as their learning sets, sweeping the field up to an entirely new level of engineering capacity.[13]

Alex Krizhevsky and his colleagues Ilya Sutskever and Geoffrey Hinton at the University of Toronto collected 1.2 million images from the web and had them crowdlabeled manually with one of 1,000 predetermined categories (crowdlabeling is made possible through websites such as Amazon Mechanical Turk; such sites harness remote labor for such needs, often from third world countries, at a cost that may be as low as $1.77 per task).[14] Training set size was the main novelty in this program, which contained eight layers in all. This program competed in a high visibility contest on September 30, 2012. It won, and its accuracy level shocked the world: image classification accuracy went up to 85%, whereas the runner-up was more than 10 percentage points behind. The leap in capacity was a landmark achievement that pioneered a new era. Predictably, the accuracy level of this program was later surpassed, as have its training set size and its number of layers. More recent machines have trained models with hundreds of layers on hundreds of millions of prelabeled items, and performance has improved. And the count keeps rising, albeit at a much slower pace, allegedly due to the law of "diminishing returns." A new era, that of *deep learning*, has begun. Let's see what effect it has had on automated language processing.

We will soon talk about the "neuron" and "synapse" metaphors and about the fruitful dialog between brain scientists and network theoreticians. But before going there, note that even the relatively small network displayed in figure 3.4 raises serious questions:

- *Size*. To approximate a real neural system, this network must be scaled up, in the number of layers and in the size of each. As nodes are massively interconnected, this increase would immediately affect the number of parameters in this network (the weights which are encoded in matrices W, V, U, T and vector S). To illustrate, figure 3.4 has five layers with four nodes each, and its weights $w_{\bullet,\bullet}$ form matrices W, V, U, T, which each have $4 \times 5 = 20$ parameters or weights; there is also vector S, which has four parameters. That's a total of 24 parameters. Keeping the same network architecture but increasing the number of nodes and number of layers just by one, we get a total of $25 \times 5 + 5 = 130$ parameters; scaling it up by an order of magnitude, namely 10-fold (40 nodes per layer; 50 layers), the number of parameters exceeds 80,000. That's a fast growth rate, and you can only imagine how easy it is to go up to the current size, which may be in the billions. Is any growth permissible, and how do such scaleups affect network performance?

- *Architecture.* Is any layer configuration possible, or is this instead constrained by computational considerations? Figure 3.4 showed a simple multilayered but two-dimensional network, which could be easily depicted on a flat screen. But how far can we go architecturally, and what does this tell us about the nature of computation and about its connection to neurobiology? These questions and many others must be discussed before these network models are taken to be scientific descriptions of human cognition. The next chapter touches on this discussion.

Taking Stock

This chapter presented the idea of a neural network and its evolution. We saw:

- An early computational perspective on single neuron activity
- The first perceptron: a network of neurons and a learning rule that does not work well
- That perceptrons with hidden units are capable of solving non–linearly separable problems
- Feedforward networks and RNNs
- The backpropagation algorithm
- The 2012 deep learning revolution

4 Dreams About Talking Machines: Current LLMs

The mentally disturbed do not employ the Principle of Scientific Parsimony: the most simple theory to explain a given set of facts. They shoot for the baroque.
—Philip K. Dick, *Valis*, 1981

The inherently contextual nature of words and sentences is at the heart of how LLMs work.
—Yann LeCun, 2022

What Do Language Machines Do?

The range of tasks that computing machinery has thus far been designed for is impressively broad. Notable among these tasks are those we all rely on in our daily toil. The tasks may be divided into three groups:

Group A: *string classification* tasks, notably (i) summarization and classification of texts, as in automated detection (and elimination) of fake news on social media, for which a division of strings into *true* and *false* ones is critical; and (ii) the establishment of logical relations between input strings that represent statements, for the purpose of reasoning (e.g., string B *follows from* string A, string A *contradicts* string B, etc.).[1]

Group B: *string generation* tasks, notably (i) machine translation, which, given a text in an input language L_{input}, outputs a text in L_{output} ($L_{\text{input}} \neq L_{\text{output}}$) that amounts to a proper translation (as judged by native speakers of both languages); (ii) question answering, which accepts an input query (let's say, a text that begins with *what*, *which*, *who*, etc. and/or ends with *?*) and, given a relevant database, outputs a proper answer (as judged by experts in the relevant domain); and (iii) next word prediction: given a context, name the next word in the stream—one that is not yet input.

Group C: *text-internal* tasks, notably coreference assignment, which links referentially dependent expressions to the correct antecedent expression in the context, whether within or across sentences. These may be pronouns (*he, his, their*), reflexives (*himself, themselves*), or other terms (*that man, the bastard*).

These are special tasks. Some have viewed the trend to build such specialized machines as detrimental, leading to a process that the late MIT computer scientist Patrick Winston (one of the greatest first-generation AI thinkers) dubbed at one time "the mechanistic Balkanization of AI": a collection of scattered communities that developed task-oriented programs. And while all these tasks are required for proper language processing, one of them, the *next word prediction* task, is presently a main benchmark for network functioning. It requires the machine to use prior knowledge for the identification of a word before it becomes known. This task is reminiscent of the 1950s fill-in-the-blank task that we've already encountered (see chapter 1), in which the participant is asked to complete the last word of a given sentence, such as *After pouring milk into her empty bowl at breakfast, Jeannie realized that an ingredient was missing and ran to the convenience store to get* ________. What does a good performer—whether human or digital—need to know and do, in order to make a good guess? I can imagine several distinct strategies, each relying on a different aspect of the utterance, part of which the incoming word is expected to be. First, the performer may try to determine that word's expected *meaning*, given what is known about the sentence and about the world (i.e., what people like Jeannie tend to put in a bowl with milk in the morning); likely, it's an edible entity, mixed with milk during breakfast, that can moreover be purchased in a convenience store. Second, the performer can try to identify the missing word's expected *grammatical form* ("part of speech"), given what is known about the structure of English sentences and what categories may follow a verb; likely, it is the name of a product or a mass noun (*Fruit Loops, cereal*), an adjective (*wild red cereal*) or perhaps a pronoun (*her*) that precedes the top candidate noun (*her cereal*). Third, she can try to calculate its expected *lexical identity*: given the counter we may have in our head that records the frequency of occurrence of each word in a context, what word is most likely to be found there? For now, I note that the natural language processing community has consistently opted almost exclusively for the third, quantitative path.

Readers may now wonder about a fourth possibility, namely, "all of the above." Wouldn't it supply the best approximation of how we humans act? Moreover, when engineering a machine, mightn't this strategy improve the system's performance? Finally, is there a reason for the primacy given to the next word prediction task? These are very good questions, which will be addressed below. But first, let's take a step backwards and look at how the field of natural language processing (NLP) got us here.

A Memory from a Distant Past

In prehistoric times, that is, before Windows and Apple's macOS, I kept wondering why I had to use a nontransparent code to communicate with my computer. It was clear that life would be much easier if I could just tell the machine in plain language what I wanted it to do and have it return a verbal answer. Sure enough, I was not the only one to wonder. Today, Generation Z has at its disposal bots like ChatGPT, thus far the most impressive incarnation of that ancient dream. The engine behind this chatbot belongs of course to the family of LLMs—multilayered, probabilistic, deep neural networks. Language tasks put an extra burden on deep neural networks, and they are therefore set apart from other computational tasks, especially visual ones like object identification and categorization, in which the input typically requires a single presentation, with no temporal unfolding. In the previous chapter, we saw how superfast computers with extreme storage capacity opened the way to programs like AlexNet, which are capable of processing huge image datasets. Here, we'll see how the special treatment that language materials require is delivered by LLMs. As we all know, these highly successful technologies have seen remarkable growth, both within and beyond the academic world, and they are the topic of this chapter.

Even the most playful of engineers build machines (especially expensive ones) for a purpose. This purpose may be practical, rather than scientific. Indeed, natural language processing research has for the most part been use-guided and has not attempted to understand why and how humans speak, what underlies our language ability, and what properties of the human mind and brain are behind it. Focusing on specific practical tasks and goals was important, as it liberated technology builders from scientific constraints.

In the case of language, this meant the construction of language machines without a grammar. But if grammar is an inherent human property, then the liberty to ignore it may incur a cost—a loss of potential scientific merits that a language machine may have. A lot, then, is at stake here. Let's keep that in mind and turn to the best and most powerful language-related technological inventions.

The Magical Effect of Letters and Numbers

Scientists and mystics have long been fascinated with the magic of letters, words, and texts. In the Middle Ages, it became customary to endow these with numerical properties. Muslim and Jewish numerologists developed mystical numerological methods that assigned values to letters, which was said to unveil "hidden" meanings lying in numerical equivalences between words, based on the value that each letter is assigned by some table. In this system, the Hebrew word DVAŠ (דְּבַשׁ=honey) is valued at 306, which is also the value of IŠA (אִשָּׁה=woman)—a numerical equivalence that supposedly underscores a female quality. (When writing this chapter, I was quite surprised to discover a multiplicity of online calculators for Hebrew numerology.)

In the modern era, a mathematical perspective took somewhat more serious directions. Early twentieth century mathematician Andrey Markov pioneered the first scientific quantitative exploration of language, with a goal that was also related to mysticism: he tried to debunk the notion "free will," as a way to argue against religion. Markov's theological musings made him launch a quantitative study of vowel sequences in an actual text—Pushkin's novel in verse *Yevgeniy Onegin*. This led to an idea currently known as a Markov chain, a tool for measuring the connection between the probabilities of occurrence of events in a sequence. Markov's invention of a new statistical tool and its application to a literary text constitute the first quantitative linguistic investigation ever.[2] He thus laid the foundation for much of the current effort to analyze texts, which consists of the application of statistical methods of ever-growing sophistication to questions about language. The torch seems to have then been taken up by George Zipf, who in another foundational study, conducted in the 1930s, counted words in large English texts and rank-ordered them (the most frequent word at the top, then the second most frequent word, etc.). He noticed an odd regularity: the most frequent word (*the*) occurs about 70,000 times in a million word text, its successor (*of*) occurs half the number of times, the third in line (*and*) has about a third of the frequency of the top word *the*, and so on. This led him to realize that the actual frequency of occurrence of a word in a text is inversely related to its rank relative to other words, which he codified in a principle, now known as Zipf's Law. While there has been little insight about the origin of this Law and while its precise formulation is debated, it is said to be obeyed by many languages, which underscores the importance of quantitative properties of human texts.[3] A related effect, concerning frequency and human behavior, was discovered soon thereafter: tasks relating to more frequent words are performed faster than those involving less frequent ones. These two developments helped to aim the limelight at quantitative properties of language materials.

Claude Shannon's eloquent presentation of *information theory* in the late 1940s set the stage for a new era in the quantitative study of language.[4] This theory is about the transmission of information over a noisy channel. It tries to quantify the degree of certainty of an encoded message being received accurately, given the noisiness of the channel it traverses. If we can successfully calculate the amount of expected error in a transmitted message, we will try to minimize it or will at least be aware of the limited reliability of the channel. For that, Shannon proposed, information needs to be quantified. Over the years, this theory has had immense impacts on a variety of scientific and engineering fields. In the context of human language, the crucial information-theoretic notion is *entropy*, a concept borrowed from thermodynamics. The entropy of a message is the amount of information that the receiver is expected to gain from it. As we move from one symbol in a message to the next, we get a clearer picture about what's coming. Information theory expresses this as a change in entropy. At the start of transmission, entropy is at its peak, with the receiver having no information; at this stage, nothing is expected, and the occurrence of any event would be surprising. Entropy decreases once transmission begins and continues to decrease as it progresses. Figure 4.1 shows this graded process: at the beginning, our ability to guess the next word (in the box) is very limited—entropy is therefore high, as the appearance of most any word would be surprising; when we move on, we become increasingly able to make a reasonable guess about the identity of the word in the box; and at the end of the transmission, the entropy of the message is minimized.

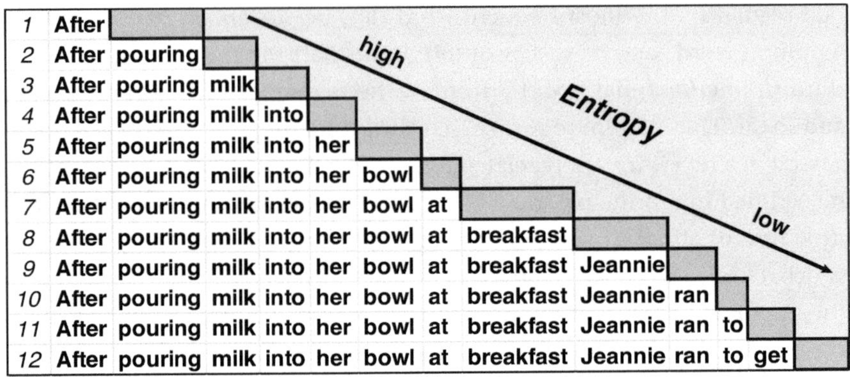

Figure 4.1
Next word prediction: a reduced version of our example *"cereal"* sentence. As the string becomes longer, the next word becomes less surprising, and easier to predict. Entropy is thus high at the beginning and decreases as the words keep unfolding.

Expected events, whose occurrence would be less surprising, thus have low entropy, while the opposite is true of unexpected events.

The relevance of information theory to communicative acts in language is quite apparent—such acts can be naturally viewed as information exchanges, aimed to reduce uncertainty. Indeed, psychologists of language immediately took notice. Most prominent among them was George Miller, the Harvard psychologist who in the early 1950s pioneered the statistical approach to language and applied information theory to comprehension (see chapter 1). This approach quickly gained prominence. However, soon thereafter, it was criticized by Noam Chomsky, who highlighted its empirical limitations, presenting language phenomena that text statistics were unable to capture. In a remarkable move, George Miller changed his mind (among the first to do so) and quickly became a leading champion of psycholinguistic studies in the new TGG-based framework, which later gained wide acceptance.

The Failure of TGG Implementations

Though TGG was based on a distinction between linguistic *competence* and *performance*, there was a (not unjustified) expectation that it could readily be implemented in a working computer program. Regarding this endeavor, the bottom line is clear: in the engineering battle with information theory, linguistic theory lost badly.

Similar to Frank Rosenblatt's colossal failure to build a working perceptron, the first language processing programs were unsuccessful: a 1950s machine translation project involving engineers, mathematicians, and linguists aimed to automate the translation of large quantities of Russian text—both published and intercepted—into English, initially with the aim of expediting the US intelligence gathering process during the Cold War (indeed, its beginnings were CIA-funded). Machine translation's goal, to build a "fully automatic high-quality translation," was never reached. A series of failures led to the formation of a special Automatic Language Processing Advisory Committee by the National Science Foundation in 1964. The committee recommended termination, funding was stopped, and the machine translation project was dead.[5] Its ultimate grave seems to have been dug by Fred Jelinek, a senior language and speech investigator at IBM Research Center, who was rumored to have snapped, "Every time I fire a linguist, the performance of the speech recognizer goes up"—a sentence that keeps being quoted by natural language processing experts. Indeed, later attempts to use symbolic grammars were not particularly successful either.

Later, I will reflect on whether the programmability, or *runnability*, of a linguistic theory is a serious test of its veracity. Yet the failures just described have clearly been held against TGG. A shift was in the making. It was perhaps most pronounced when UPenn's Mitch Marcus, the progenitor of one of the boldest computational implementations of TGG (in his MIT doctoral thesis), became a leading proponent of the statistical approach and the cocreator of the Penn Treebank, potentially the most important database for statistical computational linguistics. This database contains "approximately 7 million words of part-of-speech tagged text, 3 million words of skeletally parsed text, over 2 million words of text parsed for predicate-argument structure, and 1.6 million words of transcribed spoken text annotated for speech disfluencies."[6] The creation of the Penn Treebank laid the foundation for a new generation of statistics-based approaches, which have since prevailed in computer science (though not in linguistics), bringing us to present-day developments.

RNNs as Natural Language Generators and Analyzers

In the previous chapter, the loops in RNNs were equated to a kind of memory, as they enable previous states of the network to participate in the determination of its current state. Once recurrence is permitted, we can imagine not just one memory but *a multiplicity* of memories: one for the machine's most recent state, one representing the relative contribution of both more recent and less recent states, and so on. In the language domain, the addition of recurrence enabled these programs to relate elements in a sequence to each other during both learning and text generation. Plainly put, if past exposure gets to weigh in, learning is made easier, because it is based on a richer dataset; and text generation becomes possible once the content of your previous state (the identity of the words you've just generated) can assist in optimizing your current state. Indeed, RNNs have therefore featured in text-generating programs, mostly in question answering tasks (a.k.a. chatbots).

How does an RNN carry out its analysis of language input, and what does it require? Input comes in and is evaluated one word at a time, based on the current state of the machine, as determined by its experience with the preceding words in this string (or text). But how is a word represented as digital input for an RNN? For that, *word embedding* is a suitable tool.

What's in a Word?

When you read or hear a completely unfamiliar word, what can you do to reconstruct its meaning? When it appears in isolation, obviously you can do nothing; but when it is flanked by other words, there's a lot of information you may use in order to make an educated guess. This is, in fact, what you do when, in the middle of a conversation, somebody coughs or a car honks and the noise masks a word or two. You try to reconstruct the missing words from the context. Context may be linguistic or situational, but here, we focus on the former. The richer the context is, the more likely you are to reconstruct its missing parts correctly (as a somewhat hearing-impaired person, I do this all the time, with considerable success). From an information-theoretic perspective, the rarer the word, the lower its entropy is and the higher its informativity. Such a word is therefore more difficult to reconstruct. Shannon's idea of quantitively expressing word informativity thus fits both our intuitions and the need for a useful numerical representation of word meaning.

But how can word informativity be quantified? Based on an idea of linguist John Rupert Firth, *"You shall know a word by the company it keeps,"*[7] computational linguists invented a system of representation for each word within all its documented contexts (namely, the distribution of other words around it), in the form of a vector in a multidimensional space. The size and dimensionality of this vector, called *word embedding*, may vary depending on the nature of its designated language task. In quantitative terms, we may be able to represent word meaning through the measurement of language use parameters. This is what vector semantics tries to do. This approach quantifies each word's relevance to its context through its absolute frequency, combined with the number of times it appeared in each specific context (of other words to its left and right).

The embeddings of words that we view as similar are expected to be close to one another. Thus words that are close to one another in this vector space keep similar company, up to substitutability: that is, in a context, you may substitute one for the other, with little change in meaning. Measures of proximity are illustrated in figure 4.2, where two-valued vectors represent words along a scale of *empathy* and *intensity*. In this simple representation, the angle between the vectors and their relative length indicate how similar they are. This similarity, a word's relative frequency in the text, and the linear distance between similar words (among other parameters) all express the relevance of a particular word to others and help to establish a context for it.

Needless to say, the graph in figure 4.2 is an extremely simplified version of vector semantics, which as noted is capable of representing vectors in high-dimensional spaces. Dimensionality is determined in part by the

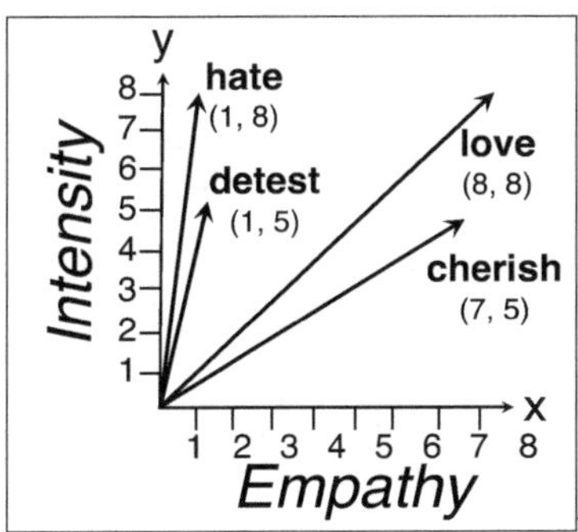

Figure 4.2
Vector-semantic representation of verbs in a two-dimensional space. The vectors for words that are close in meaning are similar.

size of the vocabulary of the training corpus we choose, normally between 10,000 and 50,000 words. But the idea is one and the same: converting meaning—both lexical and contextual—into quantities over which reliable computations can be made in the service of specific tasks. This way, Firth's oft-cited maxim is materialized.

But not all words are created equal: with grammatical words (*that, if, then, whether, whom, at least, . . .*), embedding becomes much harder. This was realized early on by philosopher Ludwig Wittgenstein, who predated Firth, and in his *Philosophical Investigations* went even further to assert that "*the meaning of a word is its use in the language.*"[8] The concepts of word embeddings and vector semantics can be thought of as implementations of his ideas as well. But even Wittgenstein qualified his statement and warned that meaning is not always just use: in his discussion of the English inflected copular verb *is*, he suggested that it has one meaning but more than one use. A broader look at the lexicon suggests that the problems actually run deeper. Still, this is how word meaning is represented in current day computational linguistics. Embeddings are the objects over which LLM's algorithms are computing, supplying the information that reduces entropy as the LLM moves along a string of words. Think again about the case where *cereal* (or *muesli*) appears in the context of a child's breakfast. When the word *breakfast* appears, the "meaning proximity" between it and *cereal*, hence the high likelihood that the latter would appear can be computed. Word embedding and vector semantics are thus critical to successful next word prediction.

How RNNs Learn with Word Embeddings

Text generation (which has lent it name to the generative AI project) is an important task that RNNs perform. The task is to deliver the next word, which is done with the help of word embeddings. For it to perform well, the RNN requires a word embedding format that is suitable to the task at hand, and a learning phase, materialized though training on a relevant training set.

After proper training as detailed in the previous chapter, the RNN starts processing its first input string, attempting to generate predictions regarding the next word. Its first prediction, of the first word, is not likely to be good: at this stage, entropy is very high, as the only information the RNN has comes from the distribution of words that are in the first position in a sentence, or more precisely, their ranking according to their likelihood to be in the first position in a sequence (their softmax distribution). We want prediction to improve as the RNN moves on the string, and so entropy has to go down. Recurrence comes to our rescue: input representing the network's state in the previous step is fed to the recurrent loop of the RNN (figure 3.5), which helps the machine to use its previous knowledge in order to narrow down the range of future possibilities. At time *t*, then, RNNs are capable of combining the word embedding at $t - 1$ into the computation, a capacity that reduces entropy and leads to improved prediction. This process is repeated until the input string reaches its end.

While recurrence is computationally intensive, RNNs are still not powerful enough for certain language tasks, because the memory that they avail a network is not sufficient for the execution of many language tasks. Next, we consider why their performance needs to be enhanced, and how this has been done.

The Need for Memory Enhancement

The discussion so far has looked at input word strings as if they have no grammatical structure, because this is how the neural network approach views language. But the intrasentential dependency relations, which were at the heart of the previous part on language, require a memory which RNNs might not be able to deliver. Salient among these are grammatical agreement relations. Consider these relations in the embedded object relative clause in 1 and the question in 2.

1. The **dogs**$_{Pl.}$ *which*^{1} *the*^{2} *girl*^{3} *showed*^{4} *her*^{5} *mother*$_{Sing.}$^{6} ~~the~~ ~~**dogs**~~ **were**$_{Pl.}$ thin

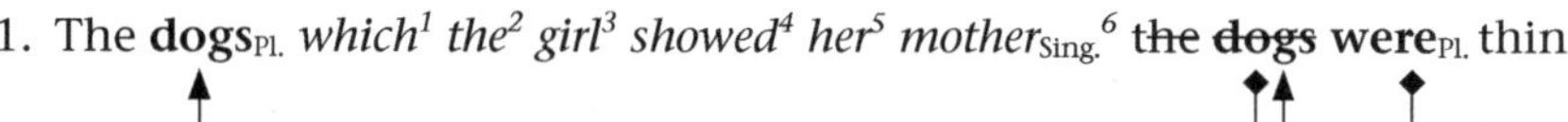

2. Which **boy**$_{Sing.}$ *did*^{1} *John*^{2} *tell*^{3} *the*^{4} *guards*$_{Pl.}$^{5} ~~which~~ ~~**boy**~~ **was**$_{Sing.}$ waiting outside

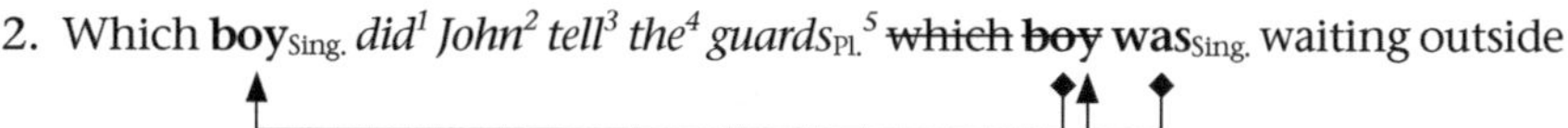

In 1, the noun phrase that is closest to the plural verb **were** is singular (*her mother*); this verb agrees with the plural **dogs** that is six words away (but its original ~~*struckthrough*~~ position is adjacent to **were**); in 2, the noun phrase closest to the singular **was** is plural (*the guards*); this verb agrees with the singular **boy** that is five words away (but its original ~~*struckthrough*~~ position is adjacent to **was**). How do these elements agree at a distance? Linguistic theory posits a MOVE relation seems to have separated the overt (heard) noun phrase and its verb. Agreement is said to hold between the original (silent, hence ~~*struckthrough*~~) position of this noun phrase and the verb (perforated line). This agreement is "inherited" by the moved noun phrase (full line).

These are frequently used sentence types, whose grammatical agreement is complex. Indeed, RNNs often fail to generate the correct forms.[9] Interestingly, much of the natural language processing literature on this matter refrains from describing these sentences in grammatical terms, sticking to the measure of linear distance between codependent positions within the string. I follow suit, and I note that from the natural language processing perspective the number of intervening words in 1 and 2 is indeed large and that the absence of grammatical analysis makes the challenge to the network even greater. As seen in 3, the number of intervening words within an intrasentential dependency relation can be even greater, in fact, almost arbitrarily increased.

3. The **boys** *who*[1] *I*[2] *introduced*[3] *to*[4] *the*[5] *girl*[6] *with*[7] *the*[8] *dragon*[9] *tattoo*[10] **were/*was** tall.

RNNs tend to focus on connections that are local, which makes correct prediction in such instances difficult, often unsuccessful. Network engineers, realizing this deficiency and being reluctant to consider the addition of grammatical dimensions into the representation, set themselves to find new solutions.

From RNNs to Long Short-Term Memory

The computational burden of enhanced recurrence is high, and when the distance between linked serial positions in a string was large, performance failures followed. A solution was proposed, to make the memory required for the computations of long-distance dependencies more efficient at expunging or at least fading out irrelevant information, while dynamically highlighting relevant elements. This device is long short-term memory (LSTM).

Instead of performing heavy computations blindly, LSTM can be thought of as managing the context through a new "memory of context" layer and the continuous assignment of new "relevance" weights to the input (through a component called an "input gate"), such that at time $t - 1$, words that are deemed to have a lesser contribution (relative to others) to the current state at time *t*, receive low weights by the "forget gate" (or even zero, in which case they are "flushed out"), while those with a larger contribution receive a greater weight (through an "output gate"). The first and last of these three components constitute the long-term aspects of the system: the input gate decides what information from state $t-1$ would be stored, and the output gate decides what information to be passed on as output. The "forget" aspect is the short-term one, hence the (intentionally oxymoronic) label *long short term memory*. LSTM gets fed with the previous state's contextual information C_{t-1}, as well as that in the current state X_t.

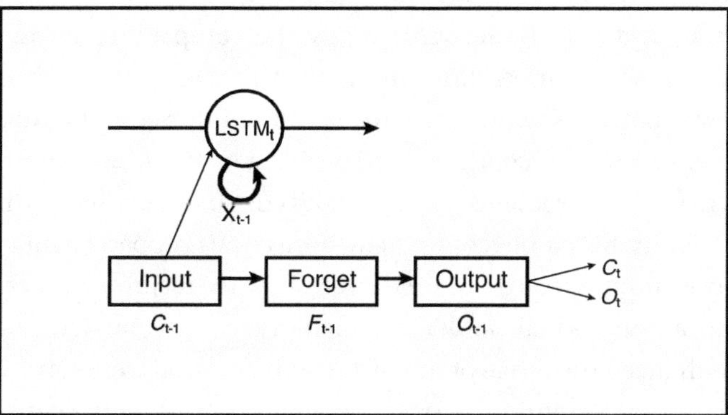

Figure 4.3
LSTM architecture that consists of input, output and forget "gates."

The decisions LSTM makes are always statistical in nature: as it is trained by the backpropagation algorithm, its weights are adjusted to identify connections in a string, such that frequent connections between nonadjacent words are highly weighted and less frequent ones can be flushed out. The result is a representation of the importance of each dependency between serial positions, which assists the network in focusing on the more frequent, hence more important, dependencies.

So where are we now? The advantage of RNNs over feedforward networks was a recurrent loop in hidden units; LSTM augmented these RNNs with a new device, namely a memory system that manages the context according to its relevance (as measured by the similarity of the embeddings of past words to the current word) and feeds the result to the output layer. Note that once again, linear distance, not structure, is what matters for

LSTM: the context is determined by cooccurrence patterns of word pairs in different serial positions in a string. Moreover, as LSTM moves along a word string, recent inputs are more relevant to the upcoming word than more distant ones. Still, the resulting memory, together with a significant increase of training data, takes us to a new level of performance in language processing, which comes with an enhanced capacity to process serial data.

Attention and Transformers: Handling Dependency Relations

We started off by talking about the distributional perspective on meaning, by which a word is to be recognized "by the company it keeps." We also discussed how word embeddings implement this idea. We then proceeded to problems that RNNs encounter in handling intrasentential dependency relations, once the codependent words are distant from one another in the string. When these problems were recognized, effort focused on increasing networks' ability to handle contextual connections properly, while reducing computation time, through the removal of costly recurrent processes. This resulted in an additional burden on memory systems, which increased in size, but enhanced the range of linguistic problems that were solved, and also increased networks' training capacity. These moves cleared the path for LLMs.

What is context? I have been using this word as a vague cover term for a variety of relations between positions in a string (that can be smaller than, equal to, or larger than a sentence). Earlier, we saw some types of dependency relations within a sentence (MOVE, pronoun-antecedent, and agreement dependencies), but there are others, and complex relations between sentences also exist. Identifying and representing these must be at the heart of any successful model for language. While linguists analyze dependency types one by one, LLM designers proposed a solution in one big sweep: a new network architecture that consists of two parts, a receiving input end ("encoder") and a transmitting output end ("decoder"). Initially designed for machine translation tasks (also known as *seq2seq* tasks) by aligning, for example, pairs of sequences from different languages, it is even capable of aligning pairs of strings that are not of equal length, which is a must in translation: expressions in different languages may mean the same thing (hence be good input–output machine translation pairs), but each may contain a different number of words, which makes their alignment and ultimately translation difficult. This enhanced encoder-decoder architecture facilitates the network's alignment task, making the establishment of dependencies easier. These capacities can be used for the establishment of dependency relations in a string, especially with the help of *attention*, a dynamic memory. Attention can represent extremely long strings, and dynamically highlight their more important aspects (similar to LSTM), so as to indicate which positions in the string should be codependent. This machine, moreover, encodes not only the text that precedes a word but also a sizeable part of the text that follows it. The result is a bidirectional machine, the *transformer*.

Let's look a bit into how a transformer processes data: the entire input string is compressed into a single vector of fixed length. This encoder vector is combined with the output of attention: for each word *w* in the string, an attention mechanism encodes all the words around *w* across the entire context (which in current LLMs can be as long as 2^{12} tokens, or several thousand English words). At each *w* position, the attention mechanism evaluates all the elements in the string along three relevance dimensions (called *query*, *key*, and *value*); the obtained values constitute an *attention score*. This score then combines with the initial string vector to produce a matrix representation, which can be thought of as a profile such that any word *w* in this string, the score highlights all those elements in the string that are of relevance to *w*. The profile of the position of *w* in the string can then be compared to that of other positions, allowing the detection of dependency relations within the string. This massive process leads to a large number of attention scores, which are now rank-ordered (by their attention score). A high score at a given position can help to detect a special relation between it and other positions. For example, in figure 4.4, each of the words *robber* and *policeman* maintains a number relation with a verb (solid line) and a gender relation with a pronoun (dotted line). These pairings of positions are therefore salient due to their high attention scores, as indicated in the figure.[10]

Transformers lack RNNs (no recurrence), which makes them more efficient, but they still require massive computation, as they link up all position pairs in a very long string. The combined information for the whole string is fed into the last part of the machine, the decoder. Still, this intense effort seems to pay off, as the encoder–decoder + attention architecture greatly enhances the performance of transformers. It fixes a deficiency that was detected in the functioning of LSTM, whose modus operandi gives precedence and saliency to recent inputs: as its most recent state encompasses a string whose latest element is the current one, a "bottleneck" is created, in which the most recent context is overrepresented, because its relevance is based on its more recent appearance than the earlier parts. The overrepresentation-of-the-recent problem has brought about performance errors. In the transformer, attention successfully ignores precedence relations of this type.[11]

(continued)

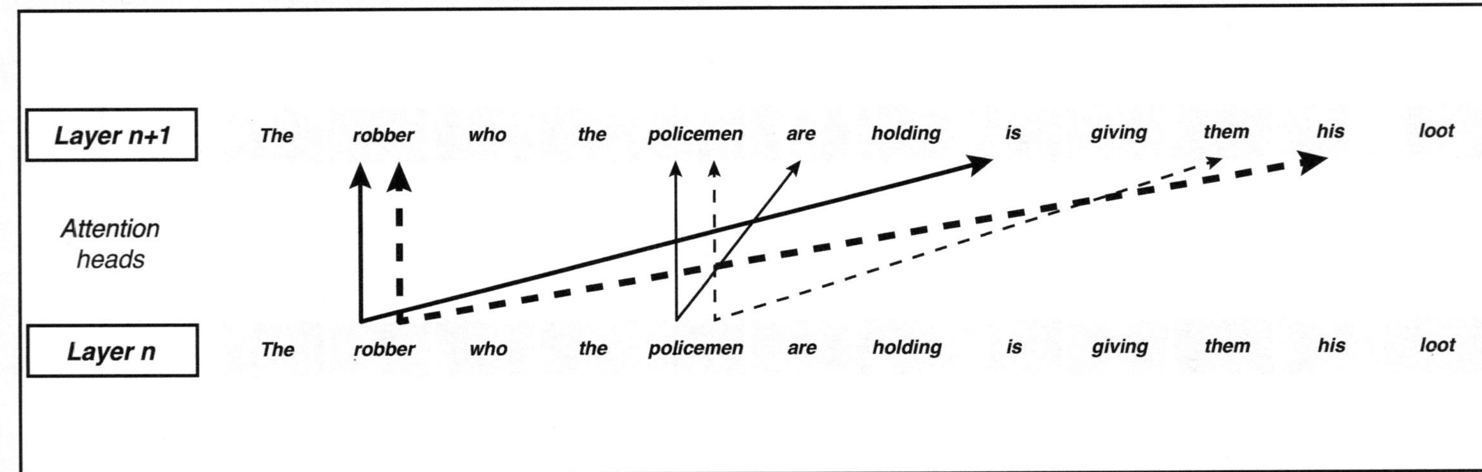

Figure 4.4
The connection that attention heads establish in a sentence that contains two agreement dependency relations.

A flagship language model using the transformer architecture has been BERT (Bidirectional Encoder Representations from Transformers)—an LLM with 110 million parameters in one version, 340 million parameters in another version.[12] Its learning phase (which I did not discuss) trains it to predict the next word. Its capacity can be scaled up by training on billions of words, affording it with more power.

Indeed, BERT's performance (as well as its offspring's) is stellar, which is what has made it extremely popular and has led to claims that it is an actual model of human language processing. This is the claim we will explore in detail in the next chapter—whether a running system with compellingly human-like performance and multiple applications for a rich variety of tasks is to be considered a scientific theory; if so, how we can know it; and how we can consider whether it meets fairly standard criteria that a valid theory of human language must satisfy.

The Interpretability of LLMs

The rapid increase in the size of neural networks has advanced their problem solving abilities but created new problems. The obvious increase in materials and energy that such gigantic systems consume, conjoined with the harmful environmental effects of this increase, has recently given rise to calls for more "frugal computing," which will likely not be heeded anytime soon.[13] In the present context, growth seems to have caused two problems. The first problem (to be discussed in the next chapter) is a nearly complete lack of constraints on LLMs' descriptive power, which sends them in a direction that is diametrically opposed to that of an explanatory theoretical science. A second problem is that our vision is blurred by opaqueness, which is due to the size of networks, their nonlinearity, the tens of billions of parameters each has (a number that's growing), and their massive connectivity. No human can conceivably approach these machines analytically, whether manually or through an interpretation algorithm. The resulting products are working machines and a lack of understanding of how and why they do what they do so well.

The problem of interpretability manifests as "the inability to effectively interpret these models," which "has debilitated their use in high-stakes applications such as medicine and raised issues related to regulatory pressure, safety, and alignment."[14] We thus end this chapter on a sweet-and-sour note, to the effect that while LLMs' operation is excellent, they are mystery black boxes. This lowers even more their chances of becoming scientific theories.

Taking Stock

This chapter updated you on present-day neural networks. We saw:

- The development of statistical methods in the analysis of texts
- The shift from structural to statistical approaches in natural language processing
- Tasks that LLMs are supposed to carry out
- RNNs
- Word embeddings and LSTM
- Attention and transformers
- The problem of interpretability

Part III Getting Under the Hood

5 A Duel: Science Versus Technology

The idea here is food that I love eating but it doesn't give me a nervous breakdown to cook. What I'm after is minimum effort and maximum pleasure, in both the cooking and the eating.

—Nigella Lawson, *Nigella Bites*, season 1, The Food Network, 2014

Chomsky did amazing things, but his time is over.

—Geoffrey Hinton, Romanes Lecture, Oxford University, 2024

Two Different Storyboards

The time has come for an evaluation, so let's wrap up. Although boundaries may be vague at times, the division of labor between science and engineering is this: engineers design and build tools, for us to use; scientists construct and test theories against observations and results, with the hope of understanding the world around us and exploring it systematically.

Engineers, who have endowed us with the gift of the new AI, with LLMs as its current top representatives in the language domain, offer a plot that's easy to follow: a brain-like model, solely based on vectorially represented text statistics, implemented in working computer programs that do things, allegedly with no symbolic rules. LLMs classify texts; engage in conversation, translation, text generation and summarization; and perform other tasks, most prominently, next word prediction. Their current makeup has rendered many objections to the old AI obsolete,[1] and a new revolution, we are told, is already here, with consequences that are not only practical but also scientific, aiming to change cognitive neuroscience. Symbolic rule systems such as those I presented in part I are obsolete.

"We now have another conceptual framework alternative," wrote AI leader Terry Sejnowski recently, "based on learning rather than writing logical computer programs."[2] Pretrained on gigantic input sets, LLMs' outputs appear human-like in a variety of tasks, which makes them a winner in any public opinion poll, including the scientific community.

Still, despite performing well, these models have not offered explanations for natural linguistic phenomena (though they may have offered explanations in other domains), nor have they been geared to discover unknown phenomena, nor have they offered more parsimonious theories (more on this later).

Linguistics as a science presents a different storyboard, one whose main aim is explanatory—to make us understand our own knowledge of language, its development, and its daily use. Indeed, it offers a theoretical apparatus whose goal is to analyze and discover our linguistic cognition. It claims that we humans possess a body of abstract yet highly constrained and precise symbolic knowledge in the form of grammatical rules, generalizations, and constraints, some of which is said to be innate; and whatever knowledge develops in the child, it does so through a rapid process, which occurs with a fairly lean input set. Language scientists formulate the shape of this knowledge and try to understand *why* it is structured the way we describe it. Common to virtually all theoreticians of virtually every linguistic color is the belief in the critical theoretical roles of symbolism, of hierarchies among symbols (at both sound and form levels), and more technically, of abstract syntactic displacement (instantiated in MOVE, for example) and the need for a semantic theory in which the meaning of a sentence is a composition of the meanings of its parts.

TGG is one such approach. It presents a theory that is grounded in formal, logic-inspired tools that give way to deterministic rules for the concatenation of symbols into meaningful strings and for the composition of their meaning representations. Probabilities play a very limited role. The claim is, moreover, that certain symbolic knowledge resides in the brain of every human baby at birth, that it is rather abstract (hence difficult to detect and characterize), that it becomes more detailed during the child's interaction with the environment during language development, and that it governs our linguistic ability as manifest in our language behavior. The reasoning behind this radical claim is the "poverty of the stimulus" argument. This class of hypotheses about linguistic knowledge and its origins is supported by a broad empirical base and multiple discoveries and is justified by solid methodological considerations. Sadly, it can take credit for few successful technological applications or accomplishments, despite multiple attempts.[3] This may be a serious weakness.

The two approaches seem incommensurate. For one thing, they disagree on the formalism (symbolic vs. probabilistic). At the same time, it is important to underscore commonalities. They give us a ray of hope for a collaborative, interesting, insightful and effective future:

- *Computationalism.* Both frameworks assume a formalism that enables computations over abstract representations.
- *Representationalism.* In both, computations are carried out over these.
- *Predictiveness.* The formal nature of the theories enables the derivation of precise predictions regarding future results.
- *Measurable biomarkers.* These computations result in measurable behavior or other precisely quantifiable biomarkers, such as varied measures of brain activity.

These commonalities are critical, as they enable a dialogue. And yet, apart from the symbolic–probabilistic difference, there are important contrasts between the two approaches, which are the topic of this chapter and the subsequent one:

A. *Innateness.* TGG takes certain grammatical knowledge to be in our head at birth. LLM proponents either deny this or blur the issue.
B. *Competence versus performance.* Chomsky's distinction between what we *actually* do and what we have *the potential* to do is a central tenet of TGG.[4] LLM proponents deny its relevance (or at least blur it).[5]
C. *Constraints.* Scientists apply various general methodological principles (e.g., parsimony, generality, transparency) to guide their exploration, and they consider the interface of their theories with other aspects of the human science (e.g., neuroscience, evolutionary sciences). In the sciences, parsimony—a theory's ability to capture more data with less theoretical machinery—is a virtue. Engineers tend to be constrained only by hardware specifications, which may force their designs to be more computationally efficient. Engineers who follow Rich Sutton's "bitter lesson" (see the epigraph to the prologue of this book) invariably prefer machines with more power, memory, speed, and computational cores to those with less.

D. *Benchmarks and empirical tests*. TGG and LLMs evaluate their systems on very different tests and use very different success criteria.[6]

Who, if anyone, is right? This chapter discusses issues A–C; the next one will address issue D from several angles. To forecast, the conclusions are that while LLM developers have given us great applications, the performance of these machines does not bode well for their status as scientific models; TGG's proponents, by contrast, have failed to build tools, but their work nonetheless constitutes a tightly constrained explanatory theory of human language, with deep insights into the mind and highly testable predictions. Let's see.

A. The Innateness Hypothesis—An Attack and Its Falsification

I have already mentioned the debate on the innateness of certain linguistic knowledge. Here is a slightly more philosophical perspective (once dubbed *Plato's problem* by Chomsky):[7] how come we know so much, given so little? That is, how does our limited linguistic knowledge empower us with the freedom to construct sentences and stories that we have never heard? Chomsky attributed this perspective on language to nineteenth century traveler and polymath Wilhelm von Humboldt, who had commented that "language is the infinite use of finite means."[8] Chomsky called this special ability "the creative aspect of language use," and he noted that we have the capacity of producing (infinitely) many sentences of arbitrary length, despite the fact that the means our head avails us are finite and that we have only been exposed to a finite set of language materials. He famously ascribed this ability to an innate "language organ" in our head, by virtue of which we have this capacity.

An Assault on Innateness and TGG

To many, the idea that we come to the world equipped with *knowledge* sounds plain crazy (though since Plato's *Meno* it has been repeatedly proposed, in different formats over the centuries). As Geoffrey Hinton recently commented, "There's this crazy guy at MIT called Chomsky, who has been claiming it's all innate. We know now it doesn't have to be the case, and Chomsky's whole view of language is kind of crazy."[9] Indeed, people tend to believe everything we know is learned, and they therefore abhor the thought of innateness. This antagonistic sentiment may have helped a recent assault on TGG to rapidly gain popularity (over 30,000 downloads in early 2025, the time of this writing). Authored by UC Berkeley's cognitive psychologist Steven Piantadosi, it is titled "Modern Language Models Refute Chomsky's Approach to Language," no less.[10] In a nutshell, its main claims are that the success of LLMs dispenses with the need for a symbolic grammar TGG-style; that LLMs "subvert and bypass the entire theoretical framework" of TGG; that the competence–performance distinction is irrelevant to an understanding of the human mind; and moreover, that in the face of LLMs' performance, the Innateness Hypothesis is falsified. This contentious paper thus touches on every item on the list of contrasts above. To bolster his claims, the author uses benchmarks that are quite different from those commended by TGG. The plan here is to examine these points by taking a careful look under the hood of the LLMs at issue.

A Quick First Falsification: LLMs Do Possess Knowledge Not Acquired on Input

From under the hood, things look a bit different. It turns out that LLMs presuppose inescapable grammatical assumptions, as they are not trained "from scratch" as is commonly claimed. Some critical tasks that LLMs are designed to perform are based on part-of speech labeling of words (and to some extent, word parts). Linguistic labels used are Noun, Verb, Adjective, and other categories we saw in earlier chapters. Obviously, these category labels do not appear in actual texts, like wiki pages, that are used for pretraining. So where does the annotation of training sets with part-of-speech labels come from? Answer: as of yet, it is not possible to realize the dream of "*You shall know a word* [only] *by the company it keeps*" in full. Therefore, an automatic annotator, or tagger, must rely, in part, on a predetermined inventory of lexical category labels for parts of speech, which is found in manually annotated—and later corrected—datasets (a.k.a. treebanks).[11] These are used at the learning stage of tagging algorithms. The part-of-speech inventory is preassigned to datasets by their curators, who decide on the tag set—the set of category labels. For example, the leading Penn Treebank has dozens of part-of-speech labels as well as syntactic tag sets for phrasal categories that its authors invented. The automatic tagging program is trained on such a manually labeled dataset and is then free to automatically annotate texts in a nearly errorless manner.[12]

Do LLMs like ChatGPT use tagged input data directly? Answer: most likely they do not (though this is a question that's a bit difficult to answer), but as far as one can tell, they all piggyback on datasets that used part-of-speech tagging in the past. Most language generation tasks and, notably, next word prediction and coreference assignment tasks rely on part-of-speech pretraining, which presupposes access to tagged input.

Take coreference assignment as an example. It links pronouns to their antecedents, as previously explained. Figure 4.4 illustrates this with the sentence *The robber who the policemen are holding is giving them his loot*, which contains two number agreement relations with auxiliary verbs (*robber–is, policemen–are*) and two pronoun–antecedent relations (*robber–his, policemen–them*). For the task to be performed successfully, agreeing verbs have to be tagged for number, and antecedent nouns, for gender, which guides the attention heads highlight them as it scans the context.[13]

Scientifically, then, there is no escape from the inevitable: a task that uses predetermined labels (whether directly or indirectly) injects the system with linguistic knowledge that did *not* come from the input. From an engineering point of view, such tagging is an efficient solution.

There may be an alternative to tagging, but it would require much more pretraining. BERT, a flagship LLM, is pretrained on a collection of texts that holds one billion words (much more than any human being is exposed to in a single lifetime)[14] and still requires some "fine-tuning" with curated, that is, tagged, data to boost its accuracy levels. I cannot even imagine the size of the input dataset that would free LLMs from this burden.

Therefore, if LLMs are a model for human language and its acquisition, their functionality cannot be said to have been built entirely on input. But in humans, knowledge acquired not on input has no way of getting into a system, and it therefore must be in it from its inception, in other words, innate. We seem to be back where we started: Piantadosi's bold statement that LLMs "subvert and bypass the entire theoretical framework" of TGG is false. Engineers may not care much about this issue, but if we are interested in science, we must be mindful of the scientific consequences of engineering choices. Skeptics may argue that this is a mere technicality that a more advanced LLM will solve. Let's wait and see; but until we have a concrete solution that we can examine, the Innateness Hypothesis is very much alive. As Gary Marcus has rightly pointed out, many Chomsky bashers have made arguments that barely scratch the surface, as the tests to which they subject LLMs are weak, at best.[15] I return to this issue in chapter 6, when I discuss an oft-heard contrarian argument, which MIT linguist Danny Fox has called the "GPT-*n* objection": *What would you say if a future version of ChatGPT gets it right?*

B. Competence and Performance: What We Do Versus What We *Can* Do

At the heart of TGG (but not LLMs) is a distinction between what speakers know about their language and what they can actually do with this knowledge. Crucial to TGG—and a major bone of contention—is its underlying claim that the linguistic *competence* we possess (the grammar we *know*) needs an implementation that turns this knowledge into action, that is, *performance* (our *use* of grammar, through our processing system). This distinction helps linguists to set up conditions to test specific grammatical rules and principles we know, while attempting to put the processing device on hold, as it were. We thus attempt to devise experiments that test the boundaries of the grammar under close-to-idealized processing conditions, in which participants (ourselves included) make judgments of plausibility, synonymy, grammaticality, and felicitous meaning and sound shape.

But in the reality of language, noise-free near-ideal test conditions are not easy to find. There are time constraints, lapses of attention, memory limitations, slips of the tongue, and other error-prone behaviors that humans exhibit in reality, which do not necessarily impinge on their knowledge of language. For example, when we hear someone say *"I've been **drunking** all night"* (a famous example given by the late UCLA linguist and speech error collector Victoria Fromkin),[16] we recognize a highly selective impairment to language mechanisms, and do not suspect that the speaker's grammatical knowledge is wiped out; rather, we attribute the error to a temporary inebriated state affecting the execution of this knowledge. Likewise, we may never be able to fully test the claim that syntactic rules can recursively build arbitrarily long sentences, because we cannot observe sentences whose number of words exceeds the number of seconds in an average human life.[17]

LLMs are not built to accommodate the competence–performance distinction. By design, they possess no grammatical knowledge (which suits the assumptions of many cognitive scientists). Indeed, we do not expect LLMs to ever make speech or language errors under conditions in which the processing device, not the grammar, is taxed. In this respect, their errorless behavior will *never* be human. Lest it be said that I am suggesting that you give your computer alcohol before you test it, let me provide a clearer, albeit somewhat more complex, example of a place where LLMs fail to simulate us and where they will keep on failing, or so it seems.

What Computers Overdo: The Embedding Paradox Speaks for the Distinction

Let me illustrate the significance of the competence–performance distinction with a sentence containing a triply embedded object relative clause (a construction that psycholinguists have been preoccupied with for years). We shall get a bit technical, but I believe that the conclusion is worth a reader's effort. In chapter 1, I discussed syntactic embedding of sentences within sentences, e.g., *John said that Bill thought that Mary loves cookies*, which is built out of 3 sentences (*John said that X, Bill thought that Y, Mary loves cookies*). Relative clauses are another type of syntactic embedding, in which a sentence describes a noun. Take S1 = ***The man fell***, S2 = ***The child pushed*** *the man*, and S3 = *The* ***girl kissed*** *the child*. You can easily understand sentence S1 when it embeds S2: *[$_{S1}$* ***The man*** *[$_{S2}$ who* ***the child pushed]*** ***fell]***. Note that the object of *pushed* is not there, as it had been displaced by MOVE). You can likewise understand sentence S2 when it embeds S3 with the same type of displacement: *[$_{S2}$* ***The child*** *[$_{S3}$ who* ***the girl kissed]*** ***pushed*** *the man]*. But as known at least since the 1950s, we cannot combine the two and expect our interlocutor to quickly understand what we are saying: the triply embedded *[$_{S1}$* ***The man*** *[$_{S2}$ who* ***the child*** *[$_{S3}$ who* ***the girl kissed] pushed] fell]*** is very difficult to understand.[18] In all cases, the embedded clause describes the subject (cf. *Which man did it? The man who the child pushed*, etc.). And yet, although the exact same recursive rule of embedding applies in all instances (NP → Noun + SG + Sentence, where SG → *who*), the result of a second application of the rule is not an acceptable English sentence (nor, certainly, is the result if you keep on embedding, as in *[$_{S1}$* ***The man*** *[$_{S2}$ who* ***the child*** *[$_{S3}$ who* ***the girl*** *[$_{S4}$ who* ***the dog bit] kissed] pushed] fell]***).

Why can certain rules of sentence embedding be applied once but result in incomprehensible sentences when applied twice or more? Do these observations speak against the embedding rule, perhaps even against recursiveness? Abolishing this rule or somehow restricting its application to a single embedding would leave the puzzle unsolved.

Moreover, here are facts that would fly in the face of this move, and lead to an Embedding Paradox: if you apply an embedding operation on the *object* of each clause (rather than embedding on the subject as above), you are free to do so many times and will still get a well-formed, easy-to-comprehend sentence, such as *[$_{S1}$ I saw* ***the man*** *[$_{S2}$ who* ***pushed the child*** *[$_{S3}$ who* ***kissed the girl*** *[$_{S4}$ who* ***bit the dog]]]]***. The subject-object contrast that the Embedding Paradox presents is important because it demonstrates that, while you are free to use recursive rules and embed as many times as you want, the freedom to embed nevertheless has an odd limit that applies only when the embedding is in the subject position of the main clause, not in the object position.

TGG resolves the paradox by distinguishing between the rules and their realization in the human head. It divides the comprehension machine into two parts: grammatical competence, which preserves the freedom to embed everywhere, and actual performance, implemented in a processing device in the human head, which has limitations that stem from architectural considerations such as working memory size. The latter is said to be responsible for the comprehension difficulty in this case. On this view, the rule applies freely, but certain resulting sentences overtax our memory and related devices, leading to a comprehension breakdown, which at times deprives us of the freedom to comprehend. Next, we turn to LLMs, to see whether their performance on the Embedding Paradox is human-like.

ChatGPT and the Embedding Paradox: Success or Failure?

LLMs are said to have no grammar. Therefore, they cannot distinguish between the grammatical freedom to embed and its processing limitations. With this in mind, I tested the Embedding Paradox with the advanced AI chatbot ChatGPT-4o—the "omni-capable" version of GPT-4—stretching it a bit with a query that contained six embeddings of this kind: "*The man whom the girl whom the boy whom the cat whom the dog whom the puppy hurt chased bit kissed pushed cried.* Who did what to whom in this sentence?" ChatGPT-4o quickly peeled the sentence like an onion, listed the actions (without assigning any grammatical structure), and concluded: "Summarizing who did what to whom in simpler terms: The puppy hurt the dog. The dog chased the cat. The cat bit the boy. The boy kissed the girl. The girl pushed the man. The man cried." Note that the full response throughout is more complex. That is, the full answer must be that *the dog* stands for *the dog whom the puppy hurt*, that *the cat* stands for *the cat whom the dog whom the puppy hurt chased*, and so one. But one should not be overly pedantic here—the machine's parsing of this complex sentence was quite good.

Why should anyone care about the fact that such sentences are parsed correctly by the LLM? They don't ever appear in texts or in conversation anyway. Am I not being overly picky? Well, if you want the LLM to exhibit human-like behavior, then here is a case where it does not—perfect comprehension behavior on sentences with multiple embedding is *not* human performance. Importantly, the Embedding Paradox seems to point to a problem that is unsolvable for LLMs: how to make the machine *weaker* in this specific case, at a time where computational power of chatbots becomes ever stronger, which will likely make the problem more severe. Note that, in the absence of a grammar-based discovery method, this problem would likely never have been detected. Should we condone an electricity-hungry industrial effort that builds such overpowerful devices (which in addition suck up trillions of gallons of cooling water from local wells)?

To borrow a phrase from Chomsky—and as visualized in figure 5.1—the LLM is a supermachine bulldozing its way through an utterance. From the perspective of syntactic embedding, it is plain superhuman. This superfluous strength should fly in the face of psychologists who jeer the competence–performance distinction and cheer ChatGPT. TGG is often mocked as being too abstract and unrealistic. And here we have an important instance of a mismatch between the machine's success and the human

Figure 5.1
A bulldozer is a machine that can lift heavy weights and push heaps of soil. Humans can also lift weights and push soil. Is the bulldozer a model for human motor skills?

failure.[19] While TGG and the distinctions it makes account for human behavior, LLMs do not.

Treat yourself to a test of ChatGPT with an even larger number of embeddings. I see no reason that the LLM's performance would go down (perhaps up to its current string length limit, which reportedly is over two thousand words).

The abolition of the competence–performance distinction is unmotivated for two additional conceptual reasons that regard abstraction:

I. Think about the addition operation in the equation $a=b+c$. For reasonably small numbers, we can calculate the sum in our heads; for larger ones, we might need a Babylonian abacus; and for yet bigger ones, a supercomputer. But the facts about execution do not change the nature of the operation. The distinction between an abstract relation and its implementation thus cannot be suppressed, as it cannot in the grammatical context.

A similar observation relates to a lesson one implicitly learns in high school physics: through noise reduction by isolation of confounding factors, one creates an idealized version of nature in which everything is removed but the factor of relevance, which can then be carefully studied. In my 10th grade physics lab, we tested Newton's laws of motion on a nearly frictionless air track like the one in figure 5.2. A body traveled on a rail. Air coming from underneath the body through small air holes in the rail minimized friction, allowing us to approximate the idealized conditions that the laws of motion require, as we measured velocity and acceleration. I

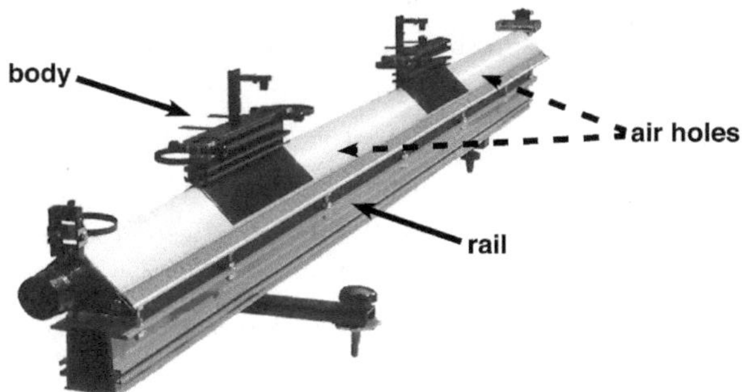

Figure 5.2
An air track that is used to test Newton's laws in high school physics laboratory experiments. Air comes through the holes and minimizes friction, but no one protests that the experimental conditions are not realistic.

do not recall any student rebelling against the idealization and arguing that Newton's laws of motion are false (or at least untestable) because they do not take friction into account. High school science teachers implant within you good principles of theory construction!

II. One of the pillars of cognitive science since the nineteenth century is the Weber–Fechner Law, which expresses the way we (and many other species) make quantity comparisons.[20] At present, some neural mechanisms that mediate it in our mind/brain are already known.[21] It was as clear to its originators then as it is to us today that the law is general, even though it requires adjustments for specific domains—each test condition and cognitive domain (whether visual, auditory, or sensorimotor) requires specific coefficients. However, it is universally accepted that these performance factors by no means diminish the generality of the law, which codifies competence in quantity comparison. It is difficult to understand why among some of those who study language, opposite positions have taken hold.

C. Constraints and Explanatory Power

What makes a theory scientific? This book is not a methodological treatise, so I will answer briefly. A description of a world can be an enumeration of the objects of that world and their properties or an abbreviated list thereof. But a scientific theory of the structure of a world explains why it is the way it is, not otherwise. A shorter description is preferrable to a longer one (provided it accounts for exactly the same phenomena) because it is more *parsimonious*; and a description that finds connections between otherwise distinct phenomena is more *general*. But a theory of these phenomena *explains* why it offers the best perspective ("inference to the best explanation"); it provides an account that tells us why the world is structured as it is, not differently. An example might elucidate these points.

The Periodic Table in Chemistry—A Severe Constraint on Possible Elements

In the mid-nineteenth century, chemists were investigating the properties of the elements through the construction of "periodic tables." Years of deliberation, combined with ingenuity, led to the periodic table high school students use to this very day, which is most strongly associated with Dmitri Mendeleev. The idea is to organize elements in a two-dimensional matrix in a manner that groups them by properties: atomic weight, metal versus nonmetal, and so on.

Mendeleev was not the first to use a table of elements. Chemists before him had arranged elements by their physical properties, with the aim of defining axes that determine the degree of proximity between related elements. But while they made lists and stopped there, he was the first to understand the table's predictive value. His table, shown in figure 5.3,

ОПЫТЪ СИСТЕМЫ ЭЛЕМЕНТОВЪ.

ОСНОВАННОЙ НА ИХЪ АТОМНОМЪ ВѢСѢ И ХИМИЧЕСКОМЪ СХОДСТВѢ.

			Ti=50	Zr=90	?=180.
			V=51	Nb=94	Ta=182.
			Cr=52	Mo=96	W=186.
			Mn=55	Rh=104,4	Pt=197,1.
			Fe=56	Rn=104,4	Ir=198.
			Ni=Co=59	Pl=106,6	Os=199.
H=1			Cu=63,4	Ag=108	Hg=200.
	Be=9,4	Mg=24	Zn=65,2	Cd=112	
	B=11	Al=27,4	?=68	Ur=116	Au=197?
	C=12	Si=28	?=70	Sn=118	
	N=14	P=31	As=75	Sb=122	Bi=210?
	O=16	S=32	Se=79,4	Te=128?	
	F=19	Cl=35,5	Br=80	I=127	
Li=7	Na=23	K=39	Rb=85,4	Cs=133	Tl=204.
		Ca=40	Sr=87,6	Ba=137	Pb=207.
		?=45	Ce=92		
		?Er=56	La=94		
		?Yt=60	Di=95		
		?In=75,6	Th=118?		

Д. Менделѣевъ

Figure 5.3
Mendeleev's original periodic table.

was rather precise and constrained, and as a result it had empty positions, which is a property that a mere list does not necessarily have.

Mendeleev noticed the empty positions and tried to understand why they were there. His brilliant idea was to realize that these gaps stood for missing elements. Each position denoted a concrete cluster of properties of unknown but chemically possible elements. That is, the empty positions not only revealed that the table was a coherent perspective on the nature of the elements that make up the universe; it was also predictive of elements that had not been found at his time. Such was the story of gallium, whose atomic weight and density Mendeleev predicted and which was isolated a mere few years later. Such too were many heavy elements, which could only be produced and detected in laboratory conditions, due to extremely short half-lives. Without a constraining periodic table, scientists would have had no guidelines that would lead to the discovery of these elements, as nothing would have told them where to look. Yet the periodic table set up clear constraints, which led them to conclude that elements in the empty slots must exist and to go hunting for them in the lab.

These stories make the periodic table easy to accept and be taught in high schools around the globe. But what is it, if not a mere abstraction? It is part of a theory of the *chemically possible*—a major step toward an understanding of the structure of the world.

Constraints on Rules and Relations in Syntax

How does Mendeleev's periodic table connect to rules of grammar? A theory of language seeks to understand the nature of linguistic knowledge, and it is therefore interested in its constitutive principles, just like the periodic table. It seeks this understanding by striving to be general and parsimonious, therefore more explanatory. Like Nigella Lawson's cooking, the linguist seeks to explain more, with less. The quest is therefore not for grammatical rules but for abstract principles that define the notion "a possible rule of grammar." Thus, a theory is explanatory if it successfully excludes all descriptive grammars on principled grounds and at the same time selects the one that describes the empirical phenomena (i.e., that is *descriptively* adequate).[22] To illustrate this rather difficult point, I will give two brief examples from TGG and then discuss the LLM approach along these lines.

Example A: X-bar theory and MERGE. In chapter 1, I introduced phrase structure rules, the rules that build basic sentence structure (e.g., Sentence → NP + VP;

VP → V+ NP; VP → V; NP → *John, Bill, Mary, cookies*, etc.). At no point did I consider why these rules and not others should be chosen or whether there are impossible phrase structure rules, that is, rules that no language could possibly have: for example, V → NP+VP; VP → N; V → *John, Bill, Mary, cookies*. These latter rule-like monsters are not part of the syntax of any human language. But while not incorporating such rules into our inventory of rules is right (in Chomsky's wording, it is right for a theory to be "descriptively adequate"), we have no understanding of *why* these monsters are excluded (which would amount to "explanatory adequacy" of the theory). For that, we need a theory of what a *humanly possible rule* is.

Returning to Mendeleev, his periodic table had the same elements as his predecessors', but in addition, there was his profound insight that "gaps" in the table were meaningful and, moreover, useful. In linguistics, a rule format called X-bar theory does that (albeit in a limited way). It provides a tightly constrained format for all phrase structure rules, one that is tailored to express the intimate connection between the term on the left of the arrow (e.g., VP in the rule VP → V+ NP) and the term(s) on its right (here: V+ NP). Details aside, this theory (currently rolled into the MERGE operation) is stated in a highly constrained and general manner, so as to enable the linguist to predict not sentences but *rules*. As gallium was a prediction of Mendeleev's approach, so does X-bar theory have the capacity to point to gaps in the X-bar format, as well as to barriers that exclude monsters such as those above.

Example B: constraints on MOVE. In chapter 2, I described MOVE and introduced a constraint on MOVE, which I called Relativized Minimality. I described transformational rules that help us to understand complex relations, for example, between a question and an answer (Q: ***Who** does Mary love?* A: *Mary loves **John***). These rules relate positions in a sentence (object position and presubject position in this case). I proceeded to discuss the constraint Relativized Minimality sets on this rule: it cannot relate two positions if an element similar to the elements in these positions intervenes between them. For example, given the answer A: *I wonder **who** could solve the problem **in this way**,* the question cannot be the ungrammatical Q: ****How** do you wonder **who** could solve this problem ~~in this way~~?* This string of words is an impossible sentence in English (and many other languages) because the MOVE relation that results in the question

(continued)

word *how* being on the left edge of the sentence crosses the question word *who*. Details aside, here, too, Relativized Minimality is a severe constraint on MOVE. Once formulated generally, Luigi Rizzi's move was to pursue the consequences of this constraint elsewhere, in surprising environments. Indeed, he was able to show that Relativized Minimality constrains the English rule of yes–no questions that was thought to be quite distinct from the question rule. Specifically (as noted in chapter 3), he discovered that the same generalization explains the contrast between the grammatical *Has the student been sailing?* And the ungrammatical **Was the student have sailing?* This connection has since become central.

Constraints, then, are at the heart of scientific theory and scientific discovery. In a sense, they provide a bird's-eye view of the scientist's daily toil and focus her attention on the best direction in which to proceed with her inquiry. Constraints on a theory also lead to testable predictions, as in the two syntactic examples I just presented.

In LLMs, It's All About More, More, *and* More

Can such constraints be found in LLMs? The answer is both *yes* and *no*. *Yes*, because LLMs are subject to hardware and run time constraints—a component that is not in keeping with the specifications of the machine on which it runs will either never get off the ground or start and crash. At times, machines are also subjected to biologically motivated learning constraints. This is true of convolutional neural networks, where interlayer connectivity is restricted, an architectural feature said to constrain the range of the learnable.[23] A possible additional example is provided by certain convolutional neural networks that deploy "constrained optimization" by setting limits on the range of image labels, as well as on other parameters.[24] There are examples of constraints that have been set up in order to fit properties of animal and human visual object recognition. In language, one finds very little of that.

The development of network architectures can be described as a sequence of additions and augmentations that improve performance, but in which language processing has until recently played a minor role. First, as I described in chapter 3, Rosenblatt proposed his perceptron as an input–output interconnected network; later, as problems were found, the concept of hidden layers was added by Minsky and Papert, who showed that such a perceptron could solve the XOR problem (which Rosenblatt's had not been able to solve); this opened the way to a host of feedforward architectures, in which connectivity and layer size could be experimented with, with no limits set on size and type; later, the idea of recurrence was added, further empowering the network. Then, networks with multiple hidden layers were coupled with AlexNet, a vast dataset of 1.2 million images collected from the web and manually categorized with one of 1,000 predetermined labels; deep networks emerged, which paved the way for the addition of LSTM and later attention components, as I described in chapter 4; finally, the quest for language modeling led to a huge increase in the capacity to represent the context, combined with multiple attention "heads" and bidirectionality, and gave us LLMs.

These are truly genius engineering developments, made possible by breathtaking, ever-growing hardware inventions. And yet, the construction of these powerful machines is not constrained by anything but what the hardware enables and the vast amounts of energy these machines require. And so, even if taken as scientific theories, LLMs fall short of explaining why the world is structured the way it is, because they are often unable to capture generalizations. Add to that the unconstrained size of sets that LLMs use for pretraining, which no human can spare enough time to process, and the humanly impossible attention mechanism, which can scan thousands of words that are stored, and you are back with Chomsky's bulldozer metaphor.

Runnability Versus Transparency and Interpretability

An oft-mentioned advantage that's attributed to LLMs is that they mostly meet a "runnability" constraint.[25] That is, they are actual programs that run on a computational platform and produce output. This property may be instrumental in making them "predictively adequate," allowing tests of novel predictions. Predictiveness of this kind is an especially important feature of devices whose internal structure is non-transparent and poorly understood. By their developers' admission, most LLMs are not "interpretable." That is, the LLM is so complex that it is not always clear which part of it is responsible to which function, which has led to a host of "interpretability" efforts, attempting to be more analytic.

Linguistic theories are not always "runnable", in that they do not easily lend themselves to be implemented as computer programs. However, these theories tend to be modular and transparent. The internal structure of most linguistic theories—and consequently, of your typical research paper in phonology, syntax, or semantics—consists of an articulation of a problem, a proposed theoretical solution, typically in the form of a set of grammatical modules that interact in ways that are relatively well understood. This structure and understanding lead to crisp predictions that are derived and reasoned, and lend themselves to a host of empirical tests. If an alternative theory exists, a comparison between it and the proposed one is in order. Thus, making a theoretical claim and then immediately deriving testable predictions is part of the daily toil of a (good) linguist.

This contrast in internal structure between the approaches could be seen earlier in this chapter, when I discussed the Embedding Paradox and pitted LLMs again the theory of syntax. The former was tested via ChatGPT-4o, whereas predictions of the latter were easily derived by looking at the structural analysis of the sentences at issue. We were able to pit the predictive adequacy of the two approaches even though only one of them was "runnable".

The runnability requirement may lead to further problems. Requiring theories to be runnable runs the risk of arbitrary, ad hoc technical assumptions concealed in the code that maps the theory onto computer action. We may call this danger *the Implementation Bias*, which pushes the quest for a running program to unintended, at times even unnoticed, ad hoc moves. Here is an example. Hebrew University's AI expert Yair Weiss and his colleagues examined the inner workings of certain programs for image identification. They discovered that tiny modifications of the method for image translation (e.g., shifting the crop by one pixel) lead to sharp performance drops in image classification by certain machines. In the cases Weiss and colleagues studied, runnability or lack thereof depends on ad hoc actions. Indeed, some of the programs they scrutinized had disappointing learning performances: a machine that was trained on 400 million images failed to learn that a tiny change in crop does not modify the image.[26] Such demonstrations point to a serious gap in LLMs' ability to generalize and transcend the specific engineering decisions that are made by designers.

In sum, a modular theory may be predictively adequate if its structure is transparent, and the functions of its pieces (or modules) can be analyzed, as illustrated by the discussion of the Embedding Paradox; at times, the eagerness to implement may lead to unintended consequences that might blur, rather than sharpen, a scientist's vision, as Weiss et al. demonstrate.

Coda: A Quick Lesson from Chess and from Weather Forecasting

An international chess scandal erupted in the summer of 2022: Magnus Carlsen, the world champion, withdrew in the middle of a game with rising star Hans Niemann, and he tweeted a link to a video clip in which a famous soccer manager can be heard saying, "I prefer really not to speak. If I speak, I am in big trouble." A subsequent investigation by chess.com resulted in a 72-page report, which concluded that Niemann "has likely cheated in more than 100 online chess games, including several prize money events."[27]

Central among the many criteria that chess.com used for the determination that Niemann had been cheating was the degree of similarity between his moves in 100 online games and those recommended by a "chess engine." Niemann's moves were too similar to the algorithm's, "strongly suggesting that he violated our fair play regulations." For chess.com, simulation-like human performance is not a sign that the program is a cognitive theory but just the opposite: it is an indication that the human is cheating, because it is clear that, unlike a scientific theory, the program is unconstrained. Indeed, when Deep Blue, in 1997, beat Gary Kasparov (figure 5.4), the then-reigning

Figure 5.4
Gary Kasparov losing to Deep Blue, 1997. Photograph by Najlah Feanny, licensed via Getty Images. © Najlah Feanny/CORBIS SABA.

world champion, no one thought Deep Blue was human. It is this event that led Noam Chomsky to comment that Deep Blue defeating Kasparov in chess was as interesting to him as a bulldozer winning the Olympics in weight lifting.

Consider now a very different research area—that of weather prediction, where LLMs outperform physics-based forecasts. For years, weather prediction has been done by a sophisticated numerical system called the Integrated Forecasting System, "which involves solving the governing equations of weather using supercomputers. . . . [Numerical weather prediction] methods are improved by highly trained experts innovating better models, algorithms, and approximations, which can be a time-consuming and costly process." A recent alternative has been machine learning–based weather prediction, "wherein forecast models are trained from historical data, including observations and analysis data."[28] It has been shown to improve forecasts by "capturing patterns in the data that are not easily represented in explicit equations," and "exploiting modern deep learning hardware, rather than supercomputers, and striking more favorable speed–accuracy trade-offs."

Weather forecasting is an area that is dedicated to prediction, not explanation. Recently, machine learning–based prediction systems have outdone traditional numerical methods which solve thousands of equations. Yet the creators of the machines understand their limitations and they emphasize that theirs is not a physical theory but an unconstrained simulation that works well: "Our approach should not be regarded as a replacement for traditional weather forecasting methods, which have been developed for decades, rigorously tested in many real-world contexts, and offer many features we have not yet explored. Rather, our work should be interpreted as evidence that [machine learning–based weather prediction] is able to meet the challenges of real-world forecasting problems and has the potential to complement and improve the current best methods."[29]

In both chess and weather prediction, then, systems are built that may outdo constrained scientific ones. Yet their designers understand that engineered but unconstrained systems may lead to prediction accuracy but not necessarily to an explanatory scientific theory.

We should therefore wonder: given that LLMs applied to natural language are unconstrained, is there a reason to think of them as scientific theories? Is it possible to characterize a component or mechanism that *cannot* be part of an LLM? For example, what exactly is the scientific status of the ad hoc, postlearning "fine-tuning" operations (i.e., the addition of new input data) in some of the LLMs to improve performance? These and other questions arise for a reader of the current LLM literature.

Still, one hopes that these models can somehow be linguistically constrained, so as to make them cognitively more realistic. In an effort to get closer to this goal, let's now move on to examine concrete problems with LLMs' performance on language tasks.

Taking Stock

This chapter was the first of two to get under LLMs' hood. We saw:

- Some assumptions that are common to all formal approaches to language
- The falsity of a claim that LLMs' stellar language behavior "refutes" the Innateness Hypothesis
- Problems that arise once the competence–performance distinction is blurred: LLMs turn out to be too strong, in exhibiting superhuman behaviors in parsing multiple embeddings "correctly," thereby failing to emulate human behavior on the Embedding Paradox
- The supreme value of constraints in science—the periodic table in chemistry
- Constraints in syntax and how they lead to deeper understanding and new discoveries
- That deep learning has few, if any, scientific constraints, which allows arbitrary decisions
- That deep learning models may meet a runnability requirement, but this requirement may only be due to their uninterpretability. Runnability moreover may lead to an Implementation Bias
- That linguistic theories are typically transparent and modular, hence interpretable, though they may not always be runnable
- That developers explicitly recognize that the predictive success of LLMs may not argue for their validity as a scientific theory—whether of chess cognition or of global weather

6 Poking Linguistic Holes in LLMs (ChatGPT Included)

DAVE: *Open the pod bay door HAL!*
HAL 9000: *I'm sorry Dave, I can't do that.*
—*2001: A Space Odyssey*, 1968

If you just have a single problem to solve, then fine, go ahead and use a neural network. But if you want to do science and understand how to choose architectures, or how to go to a new problem, you have to understand what different architectures can and cannot do.
—Marvin Minsky, *HAL's Legacy: 2001's Computer as Dream and Reality*, 1998

On Buying a Lemon

An LLM salesman shows up at your doorstep, trying to convince you that her language bot performs just the task that your business needs for growth. How do you test her product?

In several content areas, machine learning scholars have developed highly structured testing methods, which evaluate the functionality and security of their machines. *Functionality* is evaluated by measuring a program's success level on a designated task. Take the image-classifying AlexNet, in the visual domain. After training on 1.2 million images (manually categorized with one of 1,000 labels), AlexNet stunningly swept a 2012 object recognition contest, where functionality was measured by the rate of erroneous categorizing. It won, as its error rate on a widely accepted test input was exceedingly low.[1] *Security* is evaluated by a program's ability to protect itself from hostile attacks, designed to make it misclassify its input. This is typically through the insertion of subtle, perturbation-causing "adversarial examples" into the input.[2] This method measures how well a program performs after being subject to inputs whose properties may thwart correct classification. Minimal amounts of noise are introduced into the sample in a manner detected by machines but not humans and leading to misclassification. These noisy images and speech sounds look and sound similar to us but not to the program. Good testing detects such weaknesses and leads to protective measures.

How are language programs evaluated? Would success rate on any task with any input text be informative regarding functionality? What would an adversarial example for a language program be? Finally, does high success rate on a task suggest that the program is a scientific model for human language behavior? These are questions I ask in this chapter, which begins with an important anecdote and proceeds to a critical discussion of benchmarks for the language aptitude of LLMs. To forecast, I will conclude that to date, despite impressive technological successes, LLMs have not been close to being a serious scientific model of human language behavior. This is a sharp conclusion, so I'd better give you convincing arguments.

Darkness in the Absence of a Grammatical Torch: The Winograd Schema Challenge

In 1950, English mathematician and computability pioneer Alan Turing proposed what would later become the "Turing test" as a bar that a computer program claimed to simulate human behavior must pass.[3] The idea is that the test is passed when a human who maintains a dialogue with this machine is unable to tell its output from that of a real human. Though Turing presented it as a mere thought experiment, with no definitions or specific tests, and moreover understood that simulation does not necessarily amount to a theoretical explanation, his challenge has been extremely influential and still features centrally in the AI literature, including currently in discussions of ChatGPT and its human-like responses.

Yet, some voices have called for the abandonment of the Turing test, arguing that it is not a good test of human-like behavior. This test, so goes a prominent critique by University of Toronto computer scientist and philosopher Hector Levesque, relies on deception—a computer attempts to simulate human behavior, and it must either be evasive or assume a false identity, in order to pose as human. These requirements may not be indicative of humanity or human intelligence.[4]

I am not sure I fully understand the merits of Levesque's argument, as deceptive behavior presupposes intention, a mental state that I find difficult to attribute to computers. Still, as this critique has gained popularity, let's turn to his alternative proposal, which he called the Winograd schema challenge—

> a binary-choice question with these properties:
>
> - Two parties are mentioned in the question (both are males, females, objects, or groups).
> - A pronoun is used to refer to one of them ("he," "she," "it," or "they," according to the parties).
> - The question is always the same: what is the referent of the pronoun?
> - Behind the scenes, there are two *special words* for the schema. There is a slot in the schema that can be filled by either word. The correct answer depends on which special word is chosen.[5]

In an oft-mentioned instance of this schema, the two parties are *trophy* and *suitcase*, the pronoun is *it*, and the two "special words" are *big* and *small*:

1. The *trophy* did not fit in the ***suitcase*** because ***it*** was too small.

   Q: What is the referent of "it"?

   A: The suitcase.

2. The ***trophy*** did not fit in the *suitcase* because ***it*** was too big.

   Q: What is the referent of "it"?

   A: The trophy.

As the examples show, the pronoun *it* may refer to either nominal antecedent—*trophy, suitcase*. The choice of predicate in the *because* clause forces a unique antecedent for *it*. We observe an antonym-induced flip: the pronoun's antecedent changes with the substitution of an adjective for its antonym (*big, small*). The challenge is to build a machine that consistently provides a correct A to Q across specific instantiations, which are many—the case in 1 and 2 is actually an elaboration on the original one, noted in Terry Winograd's doctoral dissertation:[6]

3. The ***town councillors*** refused to give the *angry demonstrators* a permit because ***they*** feared violence.

   Q: Who feared violence?

   A: The town councillors.

4. The *town councillors* refused to give the ***angry demonstrators*** a permit because ***they*** advocated violence.

   Q: Who advocated violence?

   A: The angry demonstrators.

Here, too, the pronoun's identity flips, depending on the choice of predicate in the *because* clause—*fear, advocate*. Over the years, additional sentences with the same characteristics have been discovered, as this challenge has attracted many.[7] As no definitive solution was in sight, a decent monetary award was offered to solvers by a corporation. By 2019, the case seemed closed: "the challenge is considered defeated in 2019 since a number of transformer-based language models achieved accuracies of over 90%."[8] Indeed, if you challenge ChatGPT-4o with this task, you will surely rejoice at the machine's successful performance.

But wait: what exactly is the challenge? "Question answering" may indeed be an appropriate probe, telling us something profound about the humanity of an interlocutor; but what is it about the particular question that the Winograd schema challenge proposes that makes it so special and indicative of a machine's humanity? More generally: insightful as the challenge has been, what is so indicative about the test it uses? The idea is to find a case in which world knowledge is required for disambiguation and to show how a machine can solve it. But for disambiguation, awareness of the ambiguity itself is prerequisite. Why, then, not make the test probe this awareness directly?

To make this point clearer, let's display the ingredients of the Winograd schema challenge. It has a probe—a complex sentence that hosts a pronoun, two potential antecedents for this pronoun, and one member of an antonym pair (*big, small*; *fear, advocate*). Each member of the pair excludes a coreference relation between the pronoun and one of the two antecedents as nonsensical. The challenge is to answer a question about the identity of the antecedent, given the context.

But why is *this* the right question? If the pronoun is ambiguous, is there a reason not to test the machine's knowledge of this ambiguity directly, in contexts that allow it? This should clearly be part of the machine's ability, because it's something we humans can easily do. In the absence of a theory of language from which the ambiguity would emanate, there is no way of telling what question we must ask and why.

So here is a plan: use the Winograd schema challenge in sentences and ask the machine whether they are ambiguous, and verify that humans know they are not ambiguous (which is the reason why Levesque chose these sentences in the first place). Five years after the challenge had been declared "solved" I had the dialog shown in figure 6.1 with ChatGPT-4o.

On September 27, 2024, our favorite LLM still returns incorrect answers. I hope you realize that the important point here is *not* that ChatGPT-4o failed; in fact, as I conducted this dialogue with ChatGPT-4o several times in 2024, I can report that its replies improved over time a bit, though not entirely. Even when it did reply correctly, ChatGPT's reasoning was replete with errors once the initial sentence was truly ambiguous (e.g., *The trophy on the table seemed to not fit into the suitcase because it was crooked*). We can go on and on, and we can think creatively about other linguistically possible tests. But the problem is not solved. Note that the idea here is not to play cat and mouse with ChatGPT but rather to show that tests like the Winograd schema challenge are poor, as they lack linguistic reasoning. In syntax and semantics,

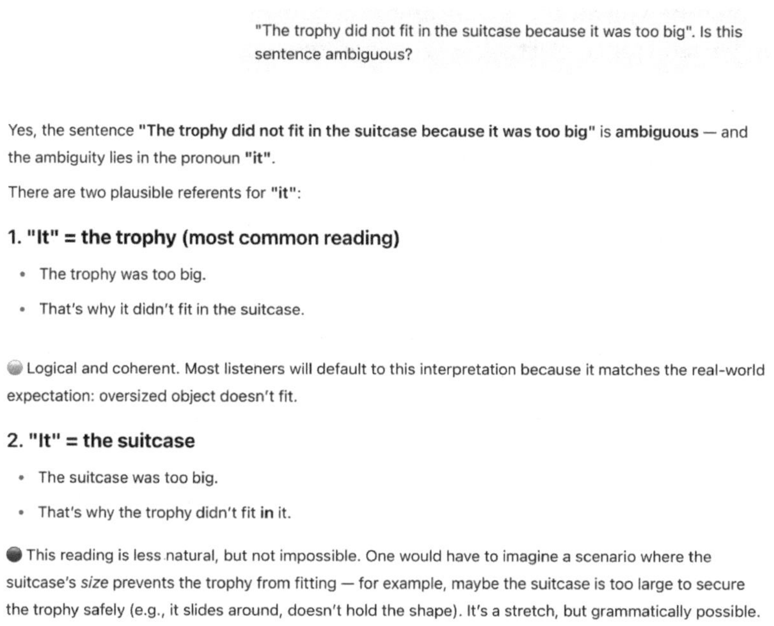

"The trophy did not fit in the suitcase because it was too big". Is this sentence ambiguous?

Yes, the sentence **"The trophy did not fit in the suitcase because it was too big"** is **ambiguous** — and the ambiguity lies in the pronoun **"it"**.

There are two plausible referents for **"it"**:

1. "It" = the trophy (most common reading)

- The trophy was too big.
- That's why it didn't fit in the suitcase.

Logical and coherent. Most listeners will default to this interpretation because it matches the real-world expectation: oversized object doesn't fit.

2. "It" = the suitcase

- The suitcase was too big.
- That's why the trophy didn't fit **in** it.

This reading is less natural, but not impossible. One would have to imagine a scenario where the suitcase's *size* prevents the trophy from fitting — for example, maybe the suitcase is too large to secure the trophy safely (e.g., it slides around, doesn't hold the shape). It's a stretch, but grammatically possible.

Figure 6.1
ChatGPT and the Winograd schema challenge (September 2024). Failure despite declarations to the contrary, apparently due to insufficient benchmarking.

pronoun–antecedent relations have long been a center of attention, with theoretical generalizations whose predictions are easily translatable into valuable tests for LLMs. But in the literature about machine learning, this scholarship is largely ignored, most unfortunately.[9] The Winograd schema challenge should thus have never been proposed as a substitute for the Turing Test (with or without a monetary prize). The problem it presents is not well-characterized, and hence no solution can be guaranteed. If the traveling salesman at your door were to sell you a Winograd schema challenge–solving bot, she'd be selling you a lemon, unbeknownst to her.

If you'd like to amuse yourself, take the vague characterization of the Winograd schema challenge, and for starters, sprinkle it with basic linguistic toys such as negation (*no, not*), quantifiers (*every, more, less*), and special words for reasoning (*because, unless, except*); add some modal predicates (*can, would, might*); and play a substitution game by putting antonym pairs (*big, small; win, lose; saint, criminal*) at the end. For example, *Every/no lawyer but John and Bill would/wouldn't take this case, because they like to win/lose. Who does the pronoun "they" refer to?* I'm sure you can take it from here, and I promise you an interesting afternoon. Just keep this in mind: unless we understand what exactly the Winograd schema challenge is, we cannot deem it "solved."

Benchmarks for Language Bots: How Informative Are They?

ChatGPT's failure on the Winograd schema challenge may be anecdotal, but it carries an important lesson: *if you don't characterize a linguistic problem correctly, you have no way of knowing whether you have solved it.* To me, the moral of this story is that, at a minimum, linguistic considerations must be a critical part of LLMs' benchmark design in the language domain. To understand how linguistic tests need to be constructed, I propose a momentary return to visual object identification. Recall that in vision, the functionality of a program is evaluated by its success rate in classifying novel images. This evaluation follows training on huge datasets of labeled images, classified into a myriad of categories that designers consider to be representative of the visual world. A program's security is tested by adversarial examples that distort images in a manner that may affect machine perception but not human perception, sneaking noise, or "malicious data." into image and audio sets.

What, if any, are the analogues of the functionality and security tests in the language domain? In the 1990s, when the Penn Treebank database was created, only functionality tests were devised. Programs were trained on half the database in a task such as part-of-speech tagging and were then tested on the other half. Life was simple. Yet at present, matters have become more complex. First, training sets have become gigantic, selected from mixed sources.[10] More importantly, part-of-speech tagging accuracy turned out to be an insufficient test. New methods for the assessment of functionality needed to be devised.

Regarding security, not much progress has been made. Compared to vision or audition, the notion of perturbations in the context of syntax and semantics is difficult to formalize (let alone automatize): in vision, as well as in music or speech, the physical properties of the signal are perturbed, creating "malicious data" that do not fool us, but fool well-designed classifiers. But at the sentence level, for ungrammatical sequences to sneak through classifiers and pretend to be grammatical, the operation must be on grammatical elements (from parts of speech on up). Hence, there is no continuous variable at the sentence level that is amenable to maneuvers that bias a machine into structured misanalysis of sentence form or meaning without being transparent to human speakers. Intentionally inserted structural ambiguity at the phrase or sentence level might work, but overall, the notion of adversarial examples in sentences and beyond needs clarification.

Indeed, I am unaware of serious attempts to construct adversarial examples or any sort of malicious data in language. The closest attempt, to my mind, has been manifested through a recent series of papers about language learning that has come out of Stanford's linguistics department. The authors introduced a series of perturbations, or "shuffles," into sentence level training sets—but they tested their LLMs' functionality, not security. The perturbations were noticeable to humans and varied in their severity. The authors measured their LLMs' speed at reaching high performance level on a next word prediction task.[11] They found that the more "shuffling" a normal English training set underwent, the more it slowed down the machine's rate of learning.

In the absence of adversarial examples that probe LLMs' security and given that next word prediction and its kin are now the central measure of success in language engineering, we may ask: Is this task optimal for the evaluation of LLMs' functionality? Scientifically speaking, does a high score on this task say much about human cognition? AI leader Ilya Sutskever answers the latter question with an emphatic affirmative: "What does it mean to predict the next word well enough? It's actually a deeper question than it seems. . . . Predicting the next token word means that you understand the underlying reality that led to the creation of that token."[12]

Sutskever may be right: predicting the next word well enough may indeed be one of several requirements of a theory that purports to make us understand the human linguistic ability. But he is not in a position to determine how good prediction is, unless he can motivate his choice of sentences for his test corpus. Next word prediction is typically carried out on texts that are either part of or similar to the training set.[13] These do not host as broad a range of sentence types as would be necessary to declare that one can obtain a "good enough" prediction of the next word. As a result, we simply do not know whether or not LLMs perform well enough, because the range of sentence types in both training and test sets is too narrow, as we saw in the discussion of the Winograd schema challenge above. Relying on the current benchmarks is therefore like testing an autonomous vehicle on a straight route with no obstacles on its path.

Upon testing on a carefully selected set of complex sentences with dependency relations, LLMs' performance on tests that feature has not been particularly successful. École Normale Supérieure and Tel Aviv University linguists Nur Lan, Emmanuel Chemla, and Roni Katzir have recently found that with simple sentences that contain dependency relations, LLMs (trained on gigantic inputs that no human ever sees in a single lifetime) predict the next word fairly well; but as soon as test sentences become a bit complex, the machines' performance drops sharply.

Much is at stake, so let me be a bit detailed here. At issue are word sequences that contain a MOVE operation and a consequent gap. Some of these sequences are grammatical sentences, but others are ungrammatical, violating constraints on MOVE (such as Relativized Minimality). The analysis of these by LLMs crucially relies on attention heads. Here, the task is next word prediction. LLMs correctly predict that *yesterday* is highly likely to appear in the sentence *I know who you talked with __ yesterday*. For this, they must expect that no noun would follow *talked with* (cf. the ungrammatical **I know who you talked with Mary yesterday*) but rather an adverb of time such as *yesterday*. Such a prediction is possible if the machine is aware that a MOVE-induced gap follows *talked with*, due to a MOVE operation that relates this position to *who*. LLMs, then, exhibit impressive performance on such a task. But how general is their ability?

On testing, the same LLMs fail to make the correct prediction when a gap appears in a somewhat more complex sentence. There are two gaps in the sentence *I know who John's talking to __ is going to annoy__soon*, annotated by "__." In these sentences, LLMs fail to predict the appearance of the adverb *soon*, indicating that they do not expect a gap after *annoy*. Lan and colleagues argue that this failure, as well as a wide variety of others they uncover, points to a severe limitation of LLMs.

By comparison, human ability to predict gaps in such complex sentences—a topic long investigated by psycholinguists—is quite good. LLMs, then, stop short of simulating our linguistic abilities once certain subtleties of natural language are introduced.

Worse yet, the same study shows that when expectations depend on the ability to invoke syntactic constraints, LLMs' next word prediction drops even further.[14]

The lesson is clear: Sutskever has no basis for declaring victory. LLMs have not proven to be a good model of human linguistic cognition. Similarly, Hinton's statement in his Nobel First Reactions interview, that "neural nets are much better at processing language than anything ever produced by the Chomsky School of Linguistics" (see prologue), is not based on a sufficiently thorough scrutiny of the linguistic capacities of LLMs.

Beyond the Winograd Schema Challenge: Extending the Range of Language Tests

Some AI experts have been trying to incorporate linguistic principles into LLM benchmarks.[15] These mostly employ classification tasks that require LLMs to mark input word strings as grammatical or ungrammatical. Such tests feature a relatively rich variety of well-formed sentence types, mixed with ungrammatical ones, whose ill-formedness stems from violations of morphological (word level), syntactic, and semantic rules and principles, in a manner illustrated in previous chapters (e.g., *cusp-s*, not **cusp-es*; *I am smart*, not *I is smarts*). LLM linguists have rightly recognized the importance and centrality of dependency relations such as MOVE and MERGE and have put them at the center of their test batteries. LLMs' performance on the ±grammaticality classification task has been relatively good, which has been taken by some as an indication that LLMs present human-like grammatical behavior and should therefore be viewed as a scientific model of human linguistic ability.

Still, the current test batteries are not as informative as one would wish them to be. Test developers face a predicament, aptly described by the authors of a test named BLiMP (Benchmark of Linguistic Minimal Pairs for English): given the richness of natural language, how do we decide where in the vast ocean of linguistic phenomena our testing should begin? How representative of human ability is our test battery? How to interpret scores that are not exactly human-like?[16] Reasonable choices of test items for the ±grammatical classification task for LLMs are marred by practical limitations, such as test duration and size, and by the lack of representativeness of the test items that are included. These limitations impose severe restrictions on the richness and subsequent informativeness of the tests that are supposed to serve as benchmarks. Coupled with the unavailability of adversity tests at the sentence level, the benchmarking of language models is difficult.

Showing deficiencies in ChatGPT has become a sport among hackers and playful linguists. I will thus be brief and point out, in the next two sections, two deficient behaviors whose significance may be broad. As the power of LLMs and the size of their training sets have grown, so has the complexity of tests that expose their deficiencies. While LLMs are an impressive technological feat that suits many practical needs and performs language-related tasks, a careful linguistic investigation uncovers its deficiencies and, more importantly, identifies the reasons for them. Two examples follow.

Drilling ChatGPT: Constraints on Pronominal Antecedents in Ellipsis

Like many natural scientists, linguists like to focus on the invisible—categories and constraints that only become apparent after complex probing and that often point to profound principles. A famous set of problems in abstract syntax, presented in chapter 2, comes from the behavior of certain ellipsis sentences. Here is a reminder:

5. ***John*** looked at ***Mary***, and so did Bill ~~*look at Mary*~~.

This complex sentence is built out of two clauses (*John looked at Mary*, *Bill looked at Mary*). A piece of the second clause is left out, yet the sentence unambiguously asserts that Bill looked at Mary, leaving no room for an alternative interpretation. Famously, though, certain ambiguities *are* allowed in ellipsis: if you throw pronouns into the pot, matters become more complicated and sensitive to extremely fine syntactic manipulations. Here is a sentence that is ambiguous in a curious way:

6. ***John*** walked his dog, *and **Mary*** and ***Bill*** did, too.

Reflect on the meaning of this sentence; of the many possibilities, it only has the following two (in the absence of a third male in the discourse, to whom *his* may have referred):

- Mary and Bill each walked his/her respective dog.
- Mary and Bill both walked John's dog.

Of the multiple referential possibilities, this sentence *cannot* mean:

- Bill walked John's dog, and Mary walked Bill's dog.
- Bill walked John's dog, and Mary walked her own dog.
- Bill walked Mary's dog, and Mary walked her own dog.
- Bill walked Mary's dog, and Mary walked John's dog.
- Bill walked Mary's dog, and Mary walked Bill's dog.

Pronouns may pick up more than one antecedent, but coreference is highly constrained. These judgments are sharp—no known interspeaker variability is associated with them. This subtlety makes such cases a good challenge for ChatGPT-4o. And indeed, it responds in a nonhuman manner, as shown in figure 6.2, indicating lack of awareness of the constraints on this sentence's meaning.

Not even the billions put in by OpenAI could beat hierarchical grammars in this case. All that training, attention, and huge memory, which no human will ever possess, led to failure. These, not the quantitative tests on large numbers of simple sentences, are the right tests, as they use extreme cases that stretch both human and machine abilities to their limit. ChatGPT collapses when required to carry out a complex comprehension task. Does this finding indicate that the machine is incapable of carrying out many tasks? Certainly not, but it does show is that the benchmarks for LLMs' performance must be linguistic.

Once again, my intention is not to play cat and mouse with ChatGPT, whose muscle keeps growing. The more important point is to realize that its abilities, which are limited at present due to a lack of syntactic and

John, Bill and Mary own different dogs. They are friends, and so they occasionally walk each other's dog, and sometimes walk a dog (or two, or three) together, but they never walk a dog that does not belong to one of them. Last Monday, John walked his dog, Bill did, too, and so did Mary. Is it possible that last Monday, John walked his own dog, Bill walked John's dog, and Mary walked her own dog?

Conclusion:

☑ **YES, it is possible** that:

- John walked his own dog (Jd),
- Bill walked John's dog (Jd),
- Mary walked her own dog (Md).

Why is this possible?

Because:

- Each "did, too" means "walked a dog" — not necessarily their own.
- Background allows sharing dogs.
- The fact that John already walked Jd does not stop Bill from also walking Jd.
- Mary walking her own dog fits perfectly well.

Figure 6.2
ChatGPT-4o fails to constrain reference assignment to pronouns in ellipsis correctly. One of the interpretations that it finds acceptable is unacceptable to native English speakers, who require a certain form of syntactic parallelism between the syntactic hierarchical structure of the missing material in the second (ellipsis) sentence and the first, full-fledged clause. This is knowledge that ChatGPT finds difficult to mimic.

semantic constraints, could be much improved if such constraints were encoded. Yet these are ignored by LLM proponents: "Because of the sheer number of homonyms and pronouns," write AI leaders Jacob Browning and Yann Lecun, "many sentences are deeply ambiguous."[17] This assertion may be true, but it is not based on evidence. In particular, it misses a central point that linguists have repeatedly underscored: that what is critical for our understanding of linguistic knowledge is not only what relations are permissible but also, perhaps more profoundly, what relations are banned. As the examples above indicate, designers' choices to ignore linguistic constraints, increasing computational power instead, lead to failure. Sadly, the reasons behind these choices seem to be sociological factors rather than engineering or scientific ones, which is detrimental to all.

A final fact—adduced by both experiment and casual observation—is that developing children understand ellipsis structures early on in childhood, starting under the age of three.[18] The ellipsis sentences they actually utter, as well as understand, are perhaps not as complex as the current examples, but children are capable of handling ellipsis after being exposed to a relatively small number of sentences during the first few years of life, a mere fraction of the size of the training sets of LLMs.

Structural Ambiguities in Brief

In chapter 1, I discussed Chomsky's early use of ambiguities such as *They are flying planes*, which are resolved once we allude to hierarchical structure (*They [are flying] planes*; *they are [flying planes]*). As LLMs pride themselves on being rule-free, they have been famous for failing to detect such ambiguities. The top currently available model, ChatGPT-4o, with its giant transformer structure, has improved and is now capable of detecting many ambiguities. However, its reliance on lexical proximities often leads to confusion. Here, instead of examples, I will give you a rough linguistic recipe, and invite you to play. Start off with the aforementioned *flying planes* sentence or its equally famous cousin *Visiting relatives can be boring* (relatives who visit versus the act of visiting one's relatives); ask ChatGPT if it is ambiguous. These are Chomsky's original examples, so their mere fame makes the ambiguity easily detected through a web search. Next, begin elaborating your example sentence in a manner that maintains its structural ambiguity and ensures that there are no disambiguating semantic cues (e.g., *Visiting relatives can be annoying if you don't really like them*; *I believe that visiting relatives can be annoying because I've had bad experiences in the past*, and so on). The more context you add (providing you carefully ensure that your additions do not rule out one of the structurally distinct meanings), the more you'll witness how the machine begins to flail and how its responses are softened (probably by local programming aimed to minimize errors in uncertain situations, which is elegantly called "fine-tuning"), leading it to avoids mistaken statements in case of uncertainty. This goes against a basic guiding intuition regarding LLMs, namely, that the inclusion of a broader context enhances the machine's performance because it decreases entropy. In these cases, we witness performance deterioration as the size of the context is increased.

Matters get even worse it you play with semantic "scope" ambiguities (*Every student failed an exam*; *I did not pass all my exams*). I do not mean to say that GPTs are bad machines. They are good, but unless you subject them to careful linguistic testing, you might be buying a lemon. As we have seen, their present performance even on low complexity linguistic examples is far from impressive, and claims that they are good models for human linguistic ability cannot be taken too seriously.

The GPT-*n* Objection

Critical commentators like me invariably encounter the GPT-*n* objection, which I briefly mentioned in chapter 5. However impressive your demonstration of an LLM failure may be, a typical response you should expect is, "Well, you surely know that soon there will be a version of GPT that gets this sentence right and responds in a human-like fashion; *what would you say then*?"

Here is what I would say. First, I would say, "Show me the money!" I would like to examine the performance of the new system by means of the tools that (psycho)linguistic theory avails us. GPT-*n*, a metaphor for the ultimate human-like language bot, should exhibit perfect human-like classification performance and, in particular, correctly classify cases that were misclassified by GPT-[*n* – 1]. In order to ascertain perfect human-like performance, we need a testing method, without which no serious evaluation of a machine's performance can be made, as the example of the Winograd schema challenge shows. The only method I know is driven by linguistic consideration.

Second, it is almost a truism that simulation does not guarantee understanding. As Turing himself wrote candidly, "May not machines carry out something which ought to be described as thinking, but which is very different from what a man does?"[19] In other words, when a machine works well, for it to become an explanation of human behavior, its builders must, at a minimum, understand why it works well. For example, they must be able to articulate the difference between GPT-[*n* – 1] and its successor GPT-*n* from which the positive change in classification behavior can be deduced. That is, they must understand how GPT-[*n* – 1] was modified so that it now covers the facts that had previously been unaccounted for. If the improved performance is due to an increase in the machine's power, for which there is no biological or behavioral justification, then we are back at Chomsky's bulldozer metaphor, because the GPT's enhanced capacities are due to superhuman mechanisms. This GPT might work, but scientifically, this cannot count as progress, because the new machine is too strong to count as a model for human behavior.

Transparency and parsimony are ancient requirements of realistic theories. The notion that grammar as a theory of human knowledge must be parsimonious as well, is far from new. In his 1951 MA thesis, Chomsky wrote, "The sole purpose of the grammar is to generate a closed body of sentences. . . . Hence [it] must be designed in such a way as to be the most efficient, economical, and elegant device generating these sentences."[20] Such considerations are often missing from the usual practice of the trade. Famous cooking guru Julia Child once said on her TV program, "If you're alone in the kitchen and you drop the lamb, you can always just pick it up. Who's going to know?" We often don't know why the machine's performance improved; unlike cooks, we need to know.

These are rather basic methodological considerations. Sadly, they do not often feature in discussions of ChatGPT's amazing language capacities. But for a classificatory success to count as a *scientific* advancement, it must be accompanied by the reasoning I just sketched. Having peeked a bit under ChatGPT's hood and having examined its truly impressive abilities, I am far from able to declare it a scientific victory.

Taking Stock

This chapter tested the details of LLMs linguistic abilities. We saw:

- The Winograd schema challenge was erroneously thought to have been met, simply because the challenge was not characterized linguistically
- Benchmarks for language bots, especially of the GPT variety, are often vague or arbitrary, hence not as informative as we would wish them to be
- Even a transformer such as GPT-4o, with its multitude of attention heads and huge span, fails on complex syntactic classifications. Such failures appear to be due to designers' lack of attention to linguistic constraints. These failures are typically ignored because the usual benchmarks stop short of probing the full complexity of natural language
- Ambiguity detection tests evince massive failures and, moreover, expose the machine's inability to generalize from an example, contrary to what it is usually praised for
- ChatGPT-*n*, a metaphoric view of the ultimate human-like language bot, is not necessarily a scientific theory. Judging by the current means of producing human-like GPTs, their structure will move them even further away from a psycho- and neurolinguistic model of human language abilities

Part IV The Brain's Language Code

7 A Mosaic of Neurolinguistic Modules

As you grow older, the brain cells connected with anxiety begin to die. . . . You start feeling better and your voice gets deeper.
—Leonard Cohen, BBC interview, 1988

We must not ignore anatomy when speaking of the physical basis of words.
—John Hughlings Jackson, FRS, neurologist, 1878

Gabby Giffords, a Story of Language Loss and of Courage

On January 8, 2011, a lone shooter attacked a small crowd in the parking lot of a shopping center in Tucson, Arizona. His target was Gabby Giffords, a first term Arizona congresswoman, who was holding a meeting with voters. A total of 18 people were shot, six of them fatally, including a nine-year-old girl and a judge. Giffords survived the attack. A bullet went into the rear part of her brain, exiting from its front left side. On its way, the bullet damaged motor and language regions, leaving her partially paralyzed and language-impaired.

Giffords's personal story is one of courage, optimism, and resilience: together with her husband, former astronaut and current US senator Mark Kelly, she has become a fierce and outspoken supporter of gun control legislation. Judging by her media appearances, she remains very active publicly more than a dozen years later, despite her limitations, which seem to be mostly in the motor and language domains.

I have not met or examined Gabby Giffords, nor have I had access to her records. But looking at her powerful public appearances (and especially, having watched the fascinating and uplifting documentary *Gabby Giffords*

Arizona Daily Star

SERVING TUCSON SINCE 1877 · SUNDAY, JANUARY 9, 2011 · REACHING 364,300 READERS

$1.50 plus tax · Outside Southern Arizona $2.50

FINAL

REP. GIFFORDS SHOT, CRITICAL

Judge Roll, girl, aide, retiree, 2 others slain

Figure 7.1
A local Arizona headline reporting the assassination attempt.

Won't Back Down),[1] it is reasonably clear to me that she suffers from a language deficit clinically known as Broca's aphasia. A relatively mild patient, her speech is somewhat nonfluent and hesitant, and she makes grammatical errors that are typical of this syndrome. Her overall cognitive state appears intact, yet sentences she produces are short and simple, rarely containing embeddings, and she has difficulty repeating sentences uttered by others. Here is an example, recorded as Giffords was rehearsing her congressional resignation speech in front of a camera: "Action! [I] gotta step down . . . people [of] Arizona . . . step down people [of] Arizona. Right now [there's] work to do . . . work to do . . . work to do, work . . . Oh." Watching her speak a decade later, she seems to have improved a lot, but she is still language-deficient, and she appears to suffer from a motor deficit to her right side as well.

To get an insight into the brain bases of language, I will take Giffords's personal story of resilience and courage as a starting point for the clinical and scientific field of neurolinguistics. In this chapter, I will tell you a bit about this field, which tries to identify and characterize language mechanisms in the human brain. It does so on the basis of clinical observations of brain-damaged patients, functional imaging in the healthy speaking brain, and intraoperative test scores of patients who undergo awake brain surgery—three research areas in which I have long been deeply involved. I believe that we can understand the present better once we know how we got here, and so I will try to present the current state of the art together with its history.

Paul Broca and His Legacy

Gabby Giffords's cluster of signs is well-recognized, known in the medical literature since French surgeon Paul Broca presented the case of the deceased Mr. Leborgne at a meeting of the Parisian *Société Anatomique* in 1861.[2] Broca had examined Leborgne, who could hardly utter a word, and taken notes on his general behavior. Subsequent to the patient's death, Broca had inspected his brain. In his presentation, Broca emphasized the selective nature of the patient's deficit, which he described as a severe impairment to his spoken language and not much else. The behavioral limitation was causally related to the "softening of the second and third convolutions of the upper part of the left frontal lobe." This led Broca to claim that he had discovered "the seat of articulated language" (*le siège de la faculté du langage articulé*) in the human brain. Importantly, Broca never dissected Leborgne's brain. Neuroanatomy at his time was not well-developed (most scientists did not even believe that the brain was composed of cells), and his characterization of the anatomy was rather opaque. Broca thus left the brain intact (it rests preserved in a Parisian museum to this day), inspected the lesion to its surface, and stopped short of assessing its precise depth or extent.

His limited empirical observations did not stop Broca from making three bold claims regarding brain structure and function, in opposition to the beliefs of most of his contemporaries: first, *distinctness*—the idea that language constitutes a separate cognitive function; second, *uniqueness*—that this function has a unique brain location, that is, a dedicated piece of neural tissue that supports it; and third, *laterality*—that language is an exclusively left-hemispheric function, from which it follows that the visually apparent anatomic symmetry between the two halves of the brain is not an indicator of functional symmetry. As Broca emphasized, patient Leborgne's loss of language was only due to damage on the left side.

Broca's contribution to science and medicine soon became widely recognized. The brain locus of this patient's lesion was later named Broca's area, and a Paris street was even named after him in 1890; see figure 7.2. Despite

Figure 7.2
A street sign for Rue Broca in Paris. Photograph courtesy of Chabe1 (Creative Commons license CC-BY-SA 4.0).

objections from many quarters, his approach to brain structure and function has guided modern neurology for over 160 years.

Pieces of Grammar in Pieces of Brain

Two left hemispheric regions, Broca's, and one now known as Wernicke's area, have long been considered the main language "areas" of the human brain. As can be seen in figure 7.3 (taken from a seminal 1970 paper, which helped Harvard neurologist Norman Geschwind revive interest in brain–language relations),[3] Broca's area (**B**) and Wernicke's area (**W**) are connected by a thick bundle of fibers.

On this classical view, inspired by an 1885 predecessor,[4] **B** is an area entrusted with the production of language and speech, **W** is in charge of language reception, **A** is a store of words (the mental lexicon), and the cable-like structure connecting **B** and **W**—a bundle of fibers known as the arcuate fasciculus—presumably relays information between these areas.

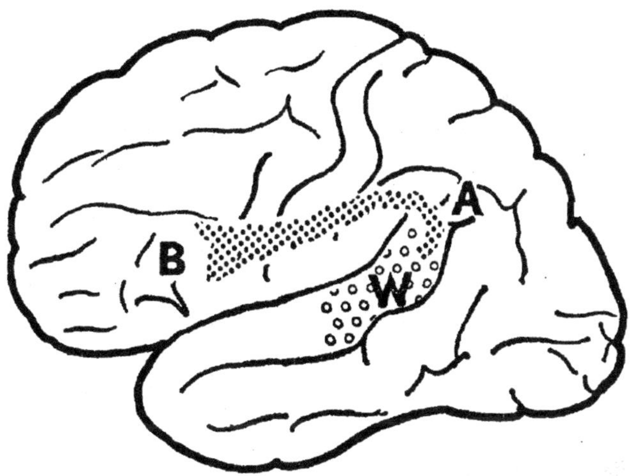

Figure 7.3
Norman Geschwind's schema of the brain's language map. Reproduced by permission from N. Geschwind, "The Organization of Language and the Brain," *Science* 170, no. 3961 (1970): 940–944.

This schema is, for all I know, the first anatomical map of a higher cognitive function ever (barring earlier phrenological drawings that were made on close to zero empirical evidence):[5] anatomical locations (large neuronal chunks) are assigned a functional role, while a thick bundle of (white matter) fibers functions as a connector cable. Indeed, this approach became known as connectionism. It was conceived on the basis of observations of patients who had suffered a focal, typically stroke-induced lesion to left Broca's or Wernicke's area, as well as to the arcuate fasciculus. Note that the patients who are most relevant to the present discussion are those with a *unifocal* brain disease; that is, the functional impairment is related to a single, identifiable, and hopefully well-delineated lesion. Patients with multiple strokes or tumors or with neurodegenerative ailments such as Alzheimer's disease do not fit this schema easily, because their pathology is either *multifocal* or *diffuse*, which stands in the way of efforts to achieve precise functional neuroanatomy.

The degree of detail of the old connectionist theory as depicted in figure 7.3 is obviously far from satisfactory. Anatomically, the extent of the named areas is at best vague, as they were marked out in an era that had no sophisticated neuroimaging devices; functionally, the terms *language production* and *language reception* refer to modalities through which language is practiced, but they make no contact with any current theoretical account of linguistic knowledge or ability. Still, Broca's, Wernicke's, and Geschwind's perspective carries an important message for cognitive neuroscientists: *pieces of cognition reside in pieces of brain.* This is neurolinguistics—the study of the functional anatomy of the network entrusted with linguistic ability in its entirety. Our task is to identify the details of the functional architecture of language and its precise brain bases. Focal brain disease is a good place to start, as it may provide important functional and anatomical evidence. Regarding the tools that may help us to discover this architecture, there are two obvious lead candidates: (psycho)linguistics on the functional side and micro- and macro-anatomy on the structural side.

Functional Modules: Our Auditory Brain Is a Phonetician

The first to understand the importance of the connection between linguistics and the brain was structural linguist Roman Jakobson, who in the 1940s started investigating patterns of phonetic breakdown in the speech of individuals with aphasia.[6] Earlier, when residing in Prague, he and his fellow Russian linguist Nikolai Trubetzkoy had proposed a theory that phonetic entities, or phonemes, are contrast-based. As we noted in chapter 1, consonants *p, t, k* are contrasted with *b, d, g* in that only the latter are "voiced" (i.e., the vocal cords start vibrating 100 milliseconds earlier than while articulating the "unvoiced" counterpart); *p, t, k* are also contrasted with *ph, th, kh* in the "manner" of their articulation (plosive vs. fricative, i.e., whether the air bursts or streams out of the mouth during articulation), whereas *p, ph* contrast with *t, th* by "place" of articulation, which is determined by the parts of the speech organ that are involved (bilabial vs. dental). A phoneme, Jakobson and Trubetzkoy said, is best represented as a cluster of abstract, hierarchically ordered contrast values, or *distinctive features* (e.g., *v*=+voiced, +fricative, +bilabial; *k*=–voiced, –fricative, +velar). Jakobson later moved to Harvard, and he and his students tried to find evidence from aphasia regarding the phonetic feature hierarchy. Their idea was that the higher a feature on the hierarchy is, the earlier it would tend to break down in aphasia.[7] They were not particularly successful in finding evidence for this idea.

Some eighty years after Jakobson and Trubetzkoy published their theory, UC San Francisco neurosurgeon Edward Chang and his team used direct cortical recording in the context of brain surgery to demonstrate compellingly that these very same abstract phonetic qualities could be directly read off the brain. This remarkable experiment involved neurosurgical patients whose brain had been equipped with a matrix of surface electrodes, placed on their auditory cortex for clinical purposes. They volunteered to participate in this experiment, in which they listened to a carefully designed human speech recording, which contained the complete phonetic inventory of English (all its consonants and vowels in different contexts). The electrodes placed on the auditory cortex of their language-dominant hemisphere directly recorded electrophysiological activity during speech perception. Since the brain produces a noisy signal, it took highly sophisticated data analysis by Nima Mesgarani to reveal a remarkable fact: certain clusters of auditory neurons respond not to the sound of a phoneme but rather to its *abstract linguistic features*; that is, speech perception involves immediate abstract linguistic analysis by our auditory cortex. Moreover, it does so along the lines of Jakobson's and Trubetzkoy's theory from the 1930s. Chang and his team read the theory directly off the patients' brains! This remarkable set of results about the linguistic neural pieces, combined with fine neuroanatomy, told a fairly clear story about the abstract nature of phonetic analysis in the human brain and its precise cerebral location.[8] This is an amazing, neurologically grounded advancement, which has since inspired many attempts to obtain clues regarding the mechanisms by which these neurons carry out abstract featural analysis.[9] If mapped anatomically, these discoveries would constitute a major step towards a neurolinguistics of the phonetic side of language.[10]

Functional Modules: Three Reasons Why Broca's Area Is a Syntax Specialist

Similar discoveries have been made about the brain bases of syntax, a topic of intense investigation and debate since the 1970s. Focusing on receptive language, I will tell you about (A) selective syntactic impairments subsequent to focal brain damage (aphasia); (B) syntactic processes as revealed through brain imaging in health (functional magnetic resonance imaging—fMRI); and (C) direct electrical stimulation of cortical tissue and concomitant syntax testing during awake brain surgery. Results from all three research areas (in which I have been personally involved for decades) indicate that our brain is divided into syntactic areas of specialty.

A. The Syntax Specialist: Errors in Poststroke Aphasia

Although expressive language impairment is apparent in individuals with Broca's aphasia, as we could sense clearly through Gabby Giffords's language patterns, testing comprehension can be more easily controlled than other language modalities, such as production, repetition, and naming. As such, comprehension has been studied extensively, with a broad range of sentence types. Controlled experimentation has revealed that the syntax of individuals with Broca's aphasia is partially impaired, evident through investigations of selective error patterns. Most striking is the fine grammatical nature of their deficit, indicating that the brain is structured by syntactic modules: in language comprehension, patients whose Broca's area has been damaged lose the ability to process grammatical transformations properly and do not lose much else.[11] Many experiments ask patients to simply indicate who did what to whom in a sentence. Even in such a simple test paradigm, a complex error pattern is observed, displaying a highly selective deficit. For example, patients interpret subject questions, like 1, correctly but miscomprehend object questions, like 2. When asked to answer the question by pointing to one of two boys in a picture, one who the girl is pushing, and the other who pushes her, the patient guesses.[12]

1. ***Which boy*** pushed the girl? *Correct performance*
2. ***Which boy*** did the girl push ~~which boy~~? *Guessing performance*

These and related results, coming from patients who are native speakers of languages as disparate as German, Hebrew, English, and Japanese, have supported the notion of a partial syntactic impairment due to a focal, stroke-induced lesion to Broca's area. They have provided important preliminary hints that neurolinguistics must incorporate grammatical principles, as these underlie the brain's modular organization for language, where certain areas specialize in highly specific linguistic functions.

Investigations of stroke victims stop short of revealing a precise anatomical basis for language. Patients with varied lesion types in different anatomical locations have been tested, only exposing a murky anatomical picture of the brain bases for language. The advent of functional neuroimaging gave new hope for more precise human functional anatomy. Imaging technologies can monitor the neural reflexes of cognitive effort in relatively small parts of the brain, which opened the way to noninvasive measurements of neural activity due to linguistic analysis (mainly in the brain's gray matter) in healthy participants. If pathological language subsequent to brain damage was at center stage until the 1990s, neurolinguistics has since gravitated toward the study of language in the intact brain, mostly through fMRI.

B. The Syntax Specialist: fMRI in Health

Spoken language comprehension necessitates all the stages in language processing: auditory, phonetic, phonological, morphological, syntactic, and semantic (in reading, visual and orthographic analyses may replace some of these). Each step taxes the brain regions that support it. The more complex the analysis, the more taxing, which is what imaging instrumentation attempts to measure. If a damaged area leads to a specific error pattern in aphasia, then in health, this same area is expected to exhibit related activation patterns. Indeed, since the mid-1990s, clues from neuroimaging studies, about the linguistic role of each brain part, have lent support to the view from aphasia and gone beyond it.[13]

A prominent early example of this effort involving the use of fMRI comes from a Max Planck Institute in Leipzig, led by neurolinguist Angela Friederici. Her team tested healthy participants' comprehension of German sentences that contained the very same words as each other but in different orders while conveying (roughly) the same meaning. German syntax allows flexibility in such sentences, and a clever neuroimaging experiment used this property to show that the more MOVE operations a sentence contains, the stronger the neural activity is in Broca's area:

3. a. Heute hat der Opa dem Jungen den Lutscher geschenkt.
   Today, Grandpa the child the lollipop gave.
   (*Today, Grandpa gave the child the lollipop.*)
   b. Heute hat ***dem Jungen*** der Opa ~~dem Jungen~~ den Lutscher geschenkt.
   1
   c. Heute hat ***dem Jungen den Lutscher*** der Opa ~~dem Jungen den Lutscher~~ geschenkt.

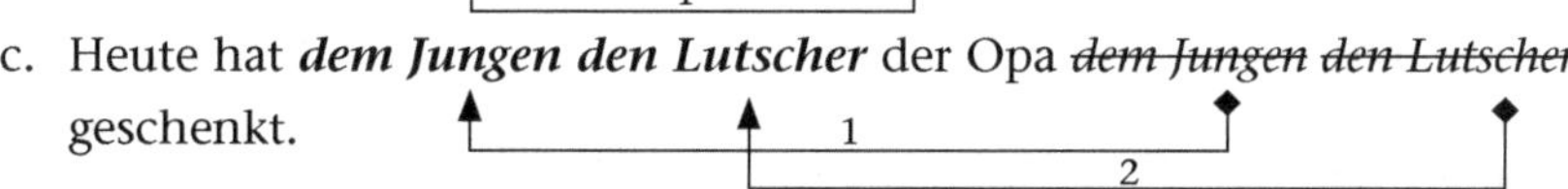

These sentences contain zero, one, and two instances of MOVE (annotated by the numbered arrows, with ~~strikethroughs~~ for the original locations of the words). The MOVE complexity increment—from 3a to 3c—determined the incremental activation measured in Broca's area (3c > 3b > 3a) and in no other brain region.

Later, neurolinguist Andrea Santi (then at my McGill laboratory in Montreal) conducted sophisticated neuroimaging experiments indicating that the increased neural activity in Broca's area reflects processing difficulty that is not merely due to the linear distance between the origin and destination of MOVE; rather, it turns out that the tissue specifically supports MOVE.[14] In one study, she showed that complexity is not determined by whether the surface order of elements is "canonical" or more frequent in the language. To this end, she kept the length of her sentences, as in 4, constant but manipulated the distance of MOVE (marked by a line) by systematically adding words between the origin and the final destination of the constituent to which MOVE applied; the expectation was that increased distance would increase the perceptual complexity of a sentence, which would lead to increased brain signal in the region that supports MOVE. She then created a set of control conditions by performing the same distance manipulation on sentences with a non-MOVE dependency relation, as in 5: she varied the length of the dependency between a reflexive pronoun *him/herself* and its antecedent noun (marked by a dashed line) by adding words between them, which increased distance.

4. a. The mailman and the mother of Jim loves ***the***

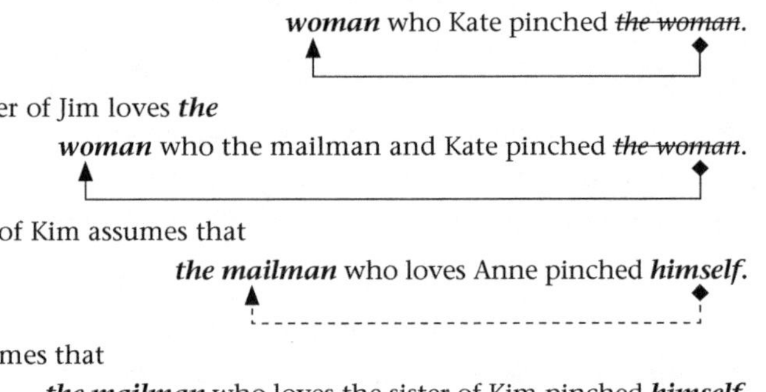

woman who Kate pinched ~~the woman~~.

b. The mother of Jim loves ***the***
woman who the mailman and Kate pinched ~~the woman~~.

5. a. The sister of Kim assumes that
the mailman who loves Anne pinched ***himself***.

b. Anne assumes that
the mailman who loves the sister of Kim pinched ***himself***.

The distance manipulation seemed a good test of MOVE selectivity in the brain: Santi reasoned that an area specializing in MOVE would activate more strongly to the added complexity in 4b versus 4a, because MOVE, which is presumably computed in Broca's area, would have higher demands in 4b (due to the greater distance between the ***bold*** and ~~*struckthrough*~~ elements). But the increased distance between the bold elements in the non-MOVE dependency in 5b versus 5a would have no impact on activation levels in Broca's area, because this dependency is not not computed in this brain part. The prediction was borne out, as a highly specific MOVE-induced activation was measured in a part of Broca's area and nowhere else in the brain. In a related study, Santi pitted MOVE complexity against embedding complexity. Once again, MOVE but not embedding activated Broca's area.[15]

The results from fMRI in health converge on those from aphasia in brain disease: those grammatical operations that are impaired by damage to Broca's area are the very same ones that activate it in health. These results reveal the functional specificity of this brain area, and the tight connection between syntactic MOVE and a specific cluster of neurons.[16]

C. The Syntax Specialist: Intraoperative Stimulation of Cortical Tissue

Further corroboration of this approach comes from a recent intraoperative study that tested the MOVE contrast with direct electrical stimulation mapping during awake brain surgery.[17] As the brain has no sensory innervation (it doesn't sense itself), patients can be awake while their brain is operated on, and intraoperative cognitive testing is sometimes surgically required. Direct electrical stimulation, used for clinical purposes, temporarily blocks a small brain area, while the awake patient is presented with a task. In this study, when Broca's area was stimulated, the comprehension of sentences with MOVE, like 6b, but not of their controls, like 6a, was transiently disrupted in several patients—they made a large number of errors on the passive sentence but hardly any on its active counterpart.

6. a. The boy is kissing the girl.
   b. ***The boy*** is kissed ~~*the boy*~~ by the girl.

Importantly, when the same stimulation procedure was repeated in other cortical loci (most notably, the left temporal lobe, which is roughly where another language region—Wernicke's area—is located), the same transient blocking did not lead to a language disruption. This highly specific error behavior lends further support to the modular, MOVE-based approach to Broca's region highlighted above.

Beyond Syntax

The fine distinctions just reported were later augmented by many others, including a connection between areas adjacent to Broca's area and highly specific and complex semantic functions.[18] For example, work in my laboratory recently uncovered the neural basis of a central logicosemantic function: negation (a.k.a. inference reversal through the insertion of words like *no, not, less, few* to a sentence). Tests we carried out in fMRI found that the contrast between sentences with and without negation selectively activates a small area in the left anterior part of the insula—a cortical structure in the frontal lobe that is somehow "buried" deeper than most cortex.[19]

Small Anatomical Modules for Language

These are mere pieces of phonetics, syntax, and semantics that constitute small functional modules. But what about their neural substrate? Can related anatomical pieces or modules be found and defined in the human brain? Can Broca's area (or parts thereof) be localized precisely? Does a function reside in a single locus or in many? A quick tour into the world of microscopic anatomy might help.

Functional anatomy requires precision. Humans feature large individual variation in brain structure (our noses, chins and foreheads are shaped differently, so why expect our brains to look alike?). To get around this variation and to provide a characterization of the brain's functional anatomy that is precise from both the function and anatomy perspectives, a method is needed that divides, or *parcellates*, the brain into coherent, function-sensitive pieces and that at the same time transcends interpersonal variability. Advances in computational neuroanatomy turn out to be very helpful here.

Our brain is famous for its complex visible surface terrain, whose convoluted topography has not been useful in dividing it into functional pieces (for one thing, about two thirds of the cortical tissue is buried in the folds, or sulci; it can only be viewed through dissection or imaged by algorithms that transform the brain into an inflated object with smooth surfaces, which entails the loss of some three-dimensional information). But the brain's cortical tissue also has rich microscopic structure, which has been explored since the late nineteenth century, an effort pioneered by Nobel laureates Santiago Ramón y Cajal and Camillo Golgi, whose work, mentioned in chapter 3, enabled the selective staining of densely organized neural tissue.[20] In the human context, Oskar and Cécile Vogt and their student Korbinian Brodmann partitioned gray matter (cortex) from white matter (fiber) and, based on the arrangement of cell layers in cortical tissue and the different cell shapes and types they contain, they parcellated the brain's cortex, or gray matter, into a mosaic of well-delineated areas. In other words, they used microscopically visible patterns that cortical tissue exhibits to build a cortical mosaic (a.k.a. "cytoarchitecture").[21]

A cytoarchitectonic area reflects local structural uniformity, which distinguishes it from its neighbors. The uniformity of a small region is believed to reflect a coherent network, from which a set of specific functions emanates. These areas are therefore the anatomical pieces that we expect to be congruent with functional components, in our case, pieces of cognitive function. Indeed, the functional relevance of the cytoarchitectonic mosaic became evident when sensorimotor and visual maps in mammals were constructed. Later, the borders of this mosaic correlated with the border of a strong marker of function—neurotransmitter receptor distribution within the cortical ribbon. A 1909 human cytoarchitectonic atlas by Brodmann consisted of 43 numbered regions (Brodmann areas—BAs). Its present-day descendants use related—though more detailed—nomenclature.[22]

The modern remapping of Broca's area (BA 44 and BA 45) by Katrin Amunts and the late Karl Zilles, of the University of Düsseldorf's Cécile and Oskar Vogt Institute and the Jülich Research Center in Germany, marked a new era in microscopic mapping of language and other abilities. It has since become part of a high resolution atlas of the human brain that divides cortical tissue into dozens of small modules, delineated through the use of microanatomical criteria. Figure 7.4 shows two adjacent cortical sections. Layers are marked (though not always visible to a nonprofessional eye), and the different cellular arrangements are apparent. Computational methods determined the boundary between the areas.[23]

Figure 7.5 displays a three-dimensional representation of the surface division of human cortex into areas in the Jülich cytoarchitectonic atlas of the human brain. Highlighted in red and blue are the parts BA 44 and BA 45, from which the slices in figure 7.4 were taken. This atlas is the best currently available method for cortical parcellation and subsequent localization of functional modules within anatomical ones.[24]

How good is the match between the anatomical and functional pieces? After all, this is at the heart of our research program, a first step in a principled relation between anatomy and function. The match can be measured by using the anatomical parcellation maps as a background to the functional localizers. To conclude that an anatomical piece is the "seat" of a functional one, we need good overlap between the anatomical (cytoarchitectonic)

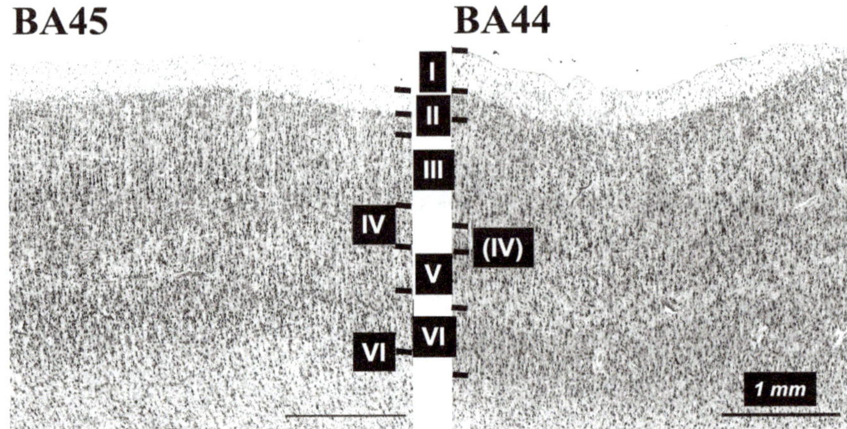

Figure 7.4
Sections from the two cytoarchitectonic parts of Broca's area, BA 44, 45. Laminar structures are different. Courtesy of Katrin Amunts, Jülich Research Center.

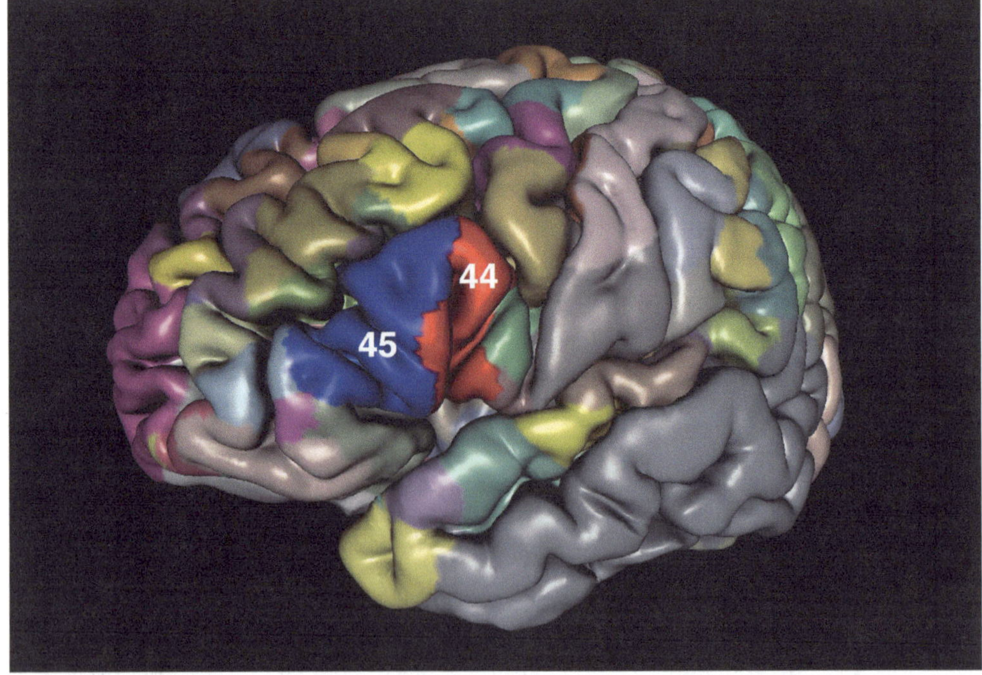

Figure 7.5
Parcellation of human cortex by cytoarchitectonic borders. BA 44, 45 are highlighted. Courtesy of Katrin Amunts, Jülich Research Center.

areas and the clusters that mark functional activation in fMRI experiments. We extract those pieces of the brain's anatomy that are known for their involvement in syntactic analysis; these are shown in figure 7.6A (red and yellow blobs). Next, we return to Santi's syntax study, which carefully isolated the brain area that supports MOVE, not other syntactic dependency relations, as well as the semantic study in my lab that isolated negation. We first asked whether Santi's activated area overlaps with a coherent piece of anatomy. We thus created a joint map, a slice of which is shown in figure 7.6B, with colored blobs from two different sources: the anatomical area, taken from 7.6A (red blob in 7.6B), and the area that was modulated by the function, drawn from Santi's fMRI results (green blob).

The image reflects the clear match between the two. The cluster of MOVE-activated neurons belongs to a single microanatomical area. The visual impression is confirmed computationally. Similar results are obtained in Santi's study on complexity, as well as in Friederici's earlier study. All three implemented the same MOVE contrast through different materials, task, controls, and language, and BA 45 kept recurring as the one area specializing in MOVE.

The negation map, discovered in my laboratory, localized to a small cortical area adjacent to Broca's area (blue blob in figure 7.6A). Again, the neural underpinnings of negation were detected, through joint work with anatomist Katrin Amunts and her Jülich team. The functional area shown in figure 7.6C was congruent with a cytoarchitectonically coherent part of the anterior insula of the left hemisphere, shown in figure 7.6D.[25] This finding provided yet another instance of the match between highly specific cognitive functions and well-defined small anatomical areas. A microscopically defined map of linguistic functions is beginning to emerge.

Is this spatial adjacency between syntactic and semantic functions accidental? I believe it is not. The "negation area" is located midway between Broca's and the anterior cingulate gyrus, where reasoning processes are known to take place.[26] As reasoning is often (though not always) done over abstract representations of linguistic content, the result suggests that the left anterior insula is a medium for these representations, which are then fed forward to the cingulate. On this interpretation of the results, the left anterior insula receives the output from Broca's area, organizes it in a logical format (i.e., like a logical formula), and it is then sent to the anterior cingulate, where reasoning operations take place. Such pathways are critical for

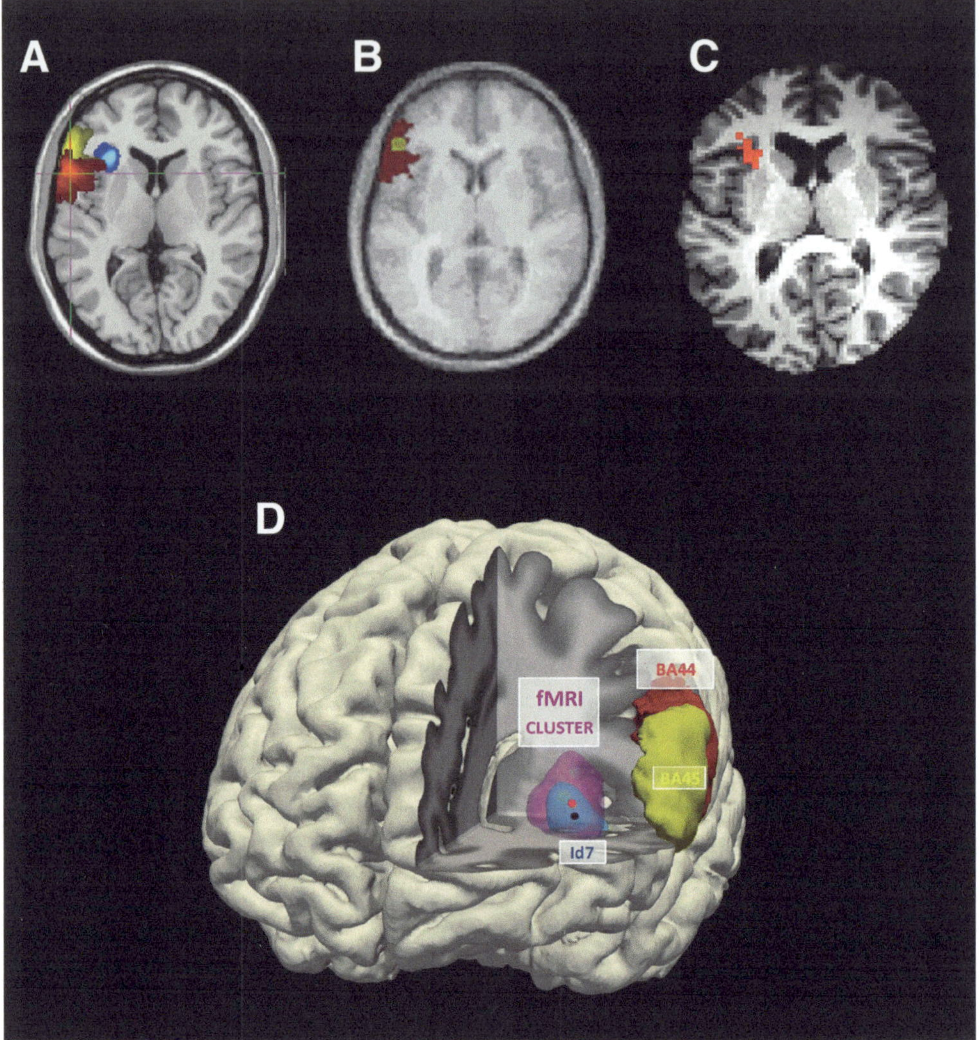

Figure 7.6

Hints regarding the microanatomy of syntax and semantics, represented in two-dimensional brain slices and a three-dimensional rendition. **A.** BA 44 (red), 45 (yellow), which together comprise Broca's area, and Id7 (blue), part of the anterior insula. All blobs are located in the left cerebral hemisphere. **B.** A typical fMRI activation cluster marking areas that support MOVE (green) superimposed on BA 45. **C.** The insular brain area that has been found to support negation. **D.** A three-dimensional depiction of the functional area in C (magenta), superimposed on the anatomically delineated area Id7. The dots represent the centers of mass of the anatomical and functional blobs, showing that they are nearly identical (difference in blob size is merely due to a "thresholding" technique). Reproduced by permission from A. Santi and Y. Grodzinsky, "Working Memory and Syntax Interact in Broca's Area," *NeuroImage* 37, no. 1 (2007): 8–17; Y. Grodzinsky, I. Deschamps, P. Pieperhoff, F. Iannilli, G. Agmon, Y. Loewenstein, and K. Amunts, "Logical Negation Mapped onto the Brain," *Brain Structure and Function* 225 (2020): 19–31.

an understanding of the neural basis of semantics and logic. Any account of the computational structure of language must include this pathway.

In sum, when we examine the current state of evidence, the neural bases for language appear to be organized as a mosaic of relatively small but connected clusters of neurons. Extant empirical findings (of which I selected just a few, simply because they associate microanatomy with microfunction) identify a clear functional-anatomical unit, one of several that have thus far been uncovered. It should be said that "small" here still translates to a large number of neurons: a quick calculation based on current data indicates that the average number of neurons we have in each of BA 44 and 45 is about 3×10^8 (three hundred million) neurons.

The approach I have described, while highly structured and logically transparent, has faced a series of objections (a putative LLM-based alternative is presented in the next chapter). Critics have claimed that regularities that seem to be observed are simply not there; that is, that the brain is more variable than depicted here.[27] The response to these objections notes that the studies they cite used linguistically unsystematic tests, which simply blurred the (otherwise clear) picture. Similar criticisms have been based on reviews of fMRI studies that claimed to find much variability in the localization of syntactic processes.[28] Are variable results due to nonuniform choice of test materials or to natural interparticipant variability?

A decisive answer seems to come from a systematic review of fMRI studies that my colleagues and I recently conducted on the neural bases of the syntactic MOVE component.[29] We scrutinized multiple fMRI studies of syntax, and selected those experiments that chose properly controlled test materials that focused on the MOVE component of sentence processing. This selection gave a clear answer, based on the brain activity detected in over 250 participants: as figure 7.7 shows, when a MOVE contrast was deployed experimentally, left Broca's area (BA 44, 45, marked red and orange) was the only brain area activated in every single study (activation points, or maxima, are superimposed on the cytoarchitecture). This was true of a variety of MOVE contrasts in English, German, and Japanese. Despite noise and false starts, then, the conjecture that specific pieces of syntax reside in designated brain pieces appears vindicated by investigations in a broad range of experiments.

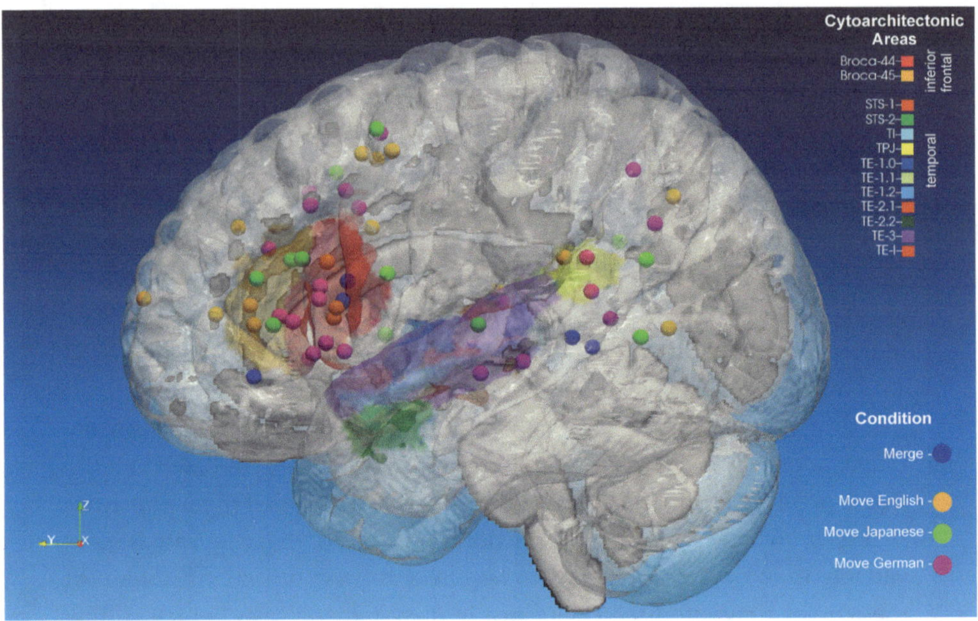

Figure 7.7
Aggregated results from 13 experiments with the MOVE contrast, superimposed on a partial cytoarchitectonic color map (see legend on right). Pins mark the maxima of MOVE activation clusters, colored by language (English = orange, Japanese = green, German = magenta). Left Broca's area (yellow and orange) was the only brain region activated in all studies. Other activation points were scattered, some in and around Wernicke's area (purple, blue and green). Such activations, however, were documented in less than half of the studies.

A Divide-and-Conquer Program and Agenda for Computational Neurolinguistics

The approach to brain–language relations that combines anatomy and linguistic theory appears vindicated. Indeed, the neuroscience of language has come a long way since Broca, Wernicke, and Geschwind: their notion of pieces of cognition in brain pieces was right, but they had no precise anatomical methods, and they missed the fact that the brain is a linguist—that its microanatomy is shaped (though perhaps not exclusively) in a manner that honors linguistic structure. Indeed, we can conclude that linguistic and microanatomical tools are crucial for the right story of the talking brain. Linguistic tools parse the function into its component parts, while cytoarchitecture parcellates cortex into functionally meaningful anatomical entities. The extant state of knowledge admittedly stops short of delivering actual neuronal models that turn structure into language functions. It lacks a description of the connectivity between areas, and it suffers technological limitations. Precise functional brain mapping in humans, especially if based on microscopic cytoarchitectonic considerations, is difficult to achieve in vivo, and can only be obtained via postmortem dissection. As functional maps require a living brain, the sampling of each individual participant for both anatomy and function is therefore not an option. Pending further technological innovations, we have to live with the fact that the resolution currently goes as high as figure 7.7 allows. Thus progress may be slow and halting, but this direction is likely the right one. Limited resolution need not limit our vision, as it did not limit early brain mappers such as McGill's neurosurgeon Wilder Penfield, who in the 1930s pioneered the mapping of the human sensorimotor cortex through the use of intraoperative direct electrical stimulation. The resulting representations, better known as *homunculi* (little persons), indicate that bodily organs are (disproportionately) represented in distinct but adjacent brain loci in both sensory and motor cortices, in a manner that later turned out to have fairly clear geometric properties.[30] Visual and auditory maps also have such structure (known as *retinotopy* and *tonotopy*) in humans as well as lower species, and we have seen precise functional (though not anatomical) maps of phonetics.

The Modularity of Language: Cytoarchitectonic Areas as a Mosaic of Linguistic Homunculi

The results presented here point to a fine mosaic that is the functional anatomy of language, made up of small, stable, and cytoarchitectonically coherent cortical substrates for major sentence processing components, discoverable only upon controlled linguistic manipulation. These small pieces, whose microanatomical makeup is known, should give rise to a detailed cortical map of linguistic mechanisms that can be modeled computationally, by anatomy-guided local networks.[31]

This program is very much in keeping with ideas that I first heard many years ago: when I was a student, a young fellow named Geoffrey Hinton came to the Center for Cognitive Science at MIT, to deliver a talk about his work on networks and parallel computation—the work for which he has just been awarded the Nobel Prize in Physics. Sitting in the audience were celebrated philosophers Daniel Dennett and his friend Jerry Fodor, who had just published his programmatic monograph *The Modularity of Mind*, in which he espoused a view of the cognitive system that divided it into functional modules, language and its components being a central one among these.[32] At the end of Hinton's talk, Fodor got up and asked (in his very special, somewhat intimidating, tone): "So, what about modularity?" Blushing, Hinton replied that his network perspective did not contradict the idea behind *The Modularity of Mind*. As far as he could tell, Hinton said, the human mind may well be composed of small modules, whose internal makeup is a set of connected networks of the kind he had presented. Fodor then said "Good!" and sat down.

I am not sure that Hinton would endorse his early position today, but his reply echoed in my mind for decades, during which I was wondering about ways to implement the modular view of the language system and of the brain. I now realize that this view is not a fantasy. The most plausible view of the linguistic brain must be that it consists of local networks that constitute small and identifiable anatomical homunculi, dictated by microscopic anatomy. These local networks power highly constrained linguistic principles and rules. As I will describe at the beginning of the next chapter, this is exactly the current practice in neurologically realistic network models of vision. The path may not be easy (and LLM investigators do not currently go there), but the evidence suggests that it is there, for us to discover.

Taking Stock

This chapter described the linguistic perspective on brain–language relations. We saw:

- The tragic yet courageous story of Gabby Giffords, who survived an assassination attempt but remained with Broca's aphasia
- Broca's, Wernicke's, and Geschwind's early language localization maps. Language is divided by the modalities through which it is practiced (speaking, listening, reading, writing)
- Distinctive phonetic features and their brain localization
- That the syntactic rule MOVE is well-localized in Broca's area, as seen in Broca's aphasia, in fMRI studies, and even through a test of MOVE in awake surgery. Other parts of syntax are located elsewhere
- Hints about semantic localization from the neuroscience of negation
- That anatomical parcellation of cortical tissue is done on the basis of microscopic definitions of cytoarchitectonic borders
- That pieces of linguistic theory and anatomical pieces are congruent in brain pieces that may contain up to 3×10^8 neurons each
- The future: more detailed neurolinguistics, better high resolution anatomy
- A modular network-based program for neurolinguistics: the quest for linguistic homunculi, akin to mapping projects in the sensorimotor systems, as well as vision and audition

8 Modeling the Linguistic Brain with LLMs

I like nonsense. It wakes up the brain cells.

—Dr. Seuss

What went wrong in the 1980s was that we all discovered that artificial intelligence technologies were actually useful. And that led to a kind of commercial distraction from which the field has never recovered.

—Patrick H. Winston, comment during panel discussion at MIT, 2011

Brain Models for Vision Can Teach Us Important Lessons

The metaphoric term *neural networks* naturally gave way to concrete implementations of real neural mechanisms. This was pioneered by visual scientists (as has often been the case), and hence vision is a good place to start; it might teach us something important that can then be applied to the modeling of language in the brain.

In 2007, Tomaso Poggio's computational neuroscience group at MIT successfully modeled early stages of primate vision with a feedforward network (that is, a network with no recurrence).[1] Predating Deep Learning, this work is based on a long tradition in the neurobiology of vision, and represents a new level of sophistication and detail, manifest in a complex model that features a "tentative mapping between the ventral stream in the primate visual system and the functional primitives of the feedforward model."

In this model (figure 8.1), the number of units within each layer varies. Its creators first showed how its early recognition of visually displayed object compares to that of human observers. Next, they measured the activation levels at the model's various layers, as evoked by objects displayed

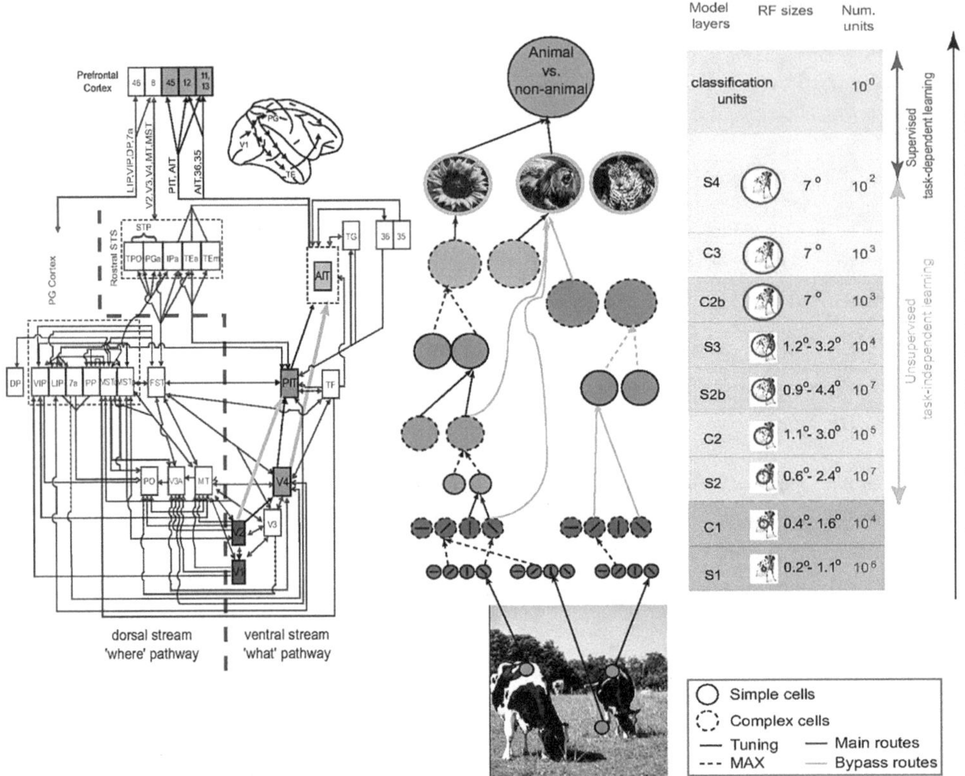

Figure 8.1
A flow chart of monkey early vision, 2007: the functional components in the monkey (bottom left), some anatomical pathways (top left), the structure of the network (middle left), and how they apply to an image (bottom middle). Reproduced by permission from T. Serre, A. Oliva, and T. Poggio, "A Feedforward Architecture Accounts for Rapid Categorization," *Proceedings of the National Academy of Sciences of the United States of America* 104 (2007): 6424–6429.

to the model. This was followed by similar experiments with primates: they measured neurophysiological results in locations of the monkey's brain that corresponded to the model's parts. Finally, they compared the model's network to the primate brain's signal. This complex sequence led the authors to conclude that their model "is closer to the anatomy and the physiology of the visual cortex in terms of quantitative parameter values" than any previously proposed ones.

The authors aptly note that the measures taken in the stations of the model fit "basic facts about the cortical mechanisms of recognition," attributing specific roles to each station and successfully simulating the functions of each of these stations. This remarkable success, coupled with other properties of the model, qualify it as a theory of the anatomicophysiological properties of the brain that produce early vision. Thus, in addition to technological advantages that such networks have proven to have, the level of empirical detail that this work provides speaks for the biological viability of its model, even if it is restricted to the earliest stages of vision. This is no small matter.

This successful construction of a network model for aspects of the visual domain, along with its validation with highly specific sets of brain data, was followed by many more, through the use of more advanced models.[2] A deep neural network–based scientific theory of visual perception and its neural bases seems to be emerging, and it seems to be composed of relatively small neuronal chunks that carry out the computation.[3]

Constitutive Principles for Neural Modeling of Perception and Cognition

Investigations of the visual system in the animal kingdom appear to be in a leading position in both experimental and theoretical neuroscience. It is their standards that of modeling we neurolinguists should seek to meet. Indeed, it seems that some principles they formulate might help us in our journey to model the language system in the brain. Several recent programmatic papers seek to articulate and implement a "neuroconnectionist" agenda.[4] They call for "explicit mapping between [artificial neural networks] and biology" for the explanation of behavior with neural networks. One of them, by Stanford philosopher/neuroscientist Rosa Cao and neuroscientist Dan Yamins, posits two general requirements for "successful explanatory models in cognitive and systems neuroscience: (a) the variables in the model correspond to components, activities, properties, and organizational features of the target mechanism . . . and (b) the dependencies posited among these variables in the model correspond to causal relations among the components of the target mechanism."

It is easy to agree with these requirements. It is equally easy to see how the feedforward network model for vision described above meets them and even goes beyond them. Its variables correspond to components of the mechanism, and there is a causal relation between these and the function;

moreover, it identifies anatomical similarity between the model's structure and direction of information flow and what is known about the brain's relevant cortical tissue, nuclei, and pathways. This feedforward network, then, *is* a scientific theory of early vision in the primate's brain.

With this in mind, let's move on to human language, where matters quickly become more complex, if not problematic.

Can LLMs Model the Linguistic Brain, and Do They?

The program for theoretical cognitive neuroscience that we just saw might help to distinguish a technological advancement from a neurobiological theory. It would therefore be wise to use it in neurolinguistics. What would the requirements spelled out by Cao and Yamins mean in the context of language? I imagine that we would need to propose neural mechanisms from which linguistic functions emanate and back up these proposals by experimental evidence of some sort. At a minimum, a theory is needed, one that would show how our structured language behavior is caused by the activity of a collection of neural mechanisms. The "pieces of language in pieces of brain" dictum would be cashed out with detailed mechanisms that causally explain linguistic functioning via an appeal to neural networks. I believe wholeheartedly in this program and am hopeful about its success; but I am not sure that we are that close to realizing it. Nevertheless, the success of LLMs has led to attempts to apply them to neurolinguistics, with the expressed hope of converging on a mechanistic neural model for language. The previous chapter showed that the path towards such a model is long and tortuous, but the excitement about LLMs has motivated some to take the risk. Here, I focus on two ambitious projects that have recently applied LLMs to the study of language mechanisms in the human brain, in an attempt to see whether they qualify as scientific theories of language or at least the beginnings thereof.

Two recent series of works on LLM–brain–language relations have received much attention, and I will briefly describe and then critique them. The first, by Princeton's Ariel Goldstein and Uri Hasson, compares aspects of the language performance of various LLMs to human performance and plainly argues that "the human brain processes incoming speech similarly to an autoregressive [deep language model]"[5] and that Goldstein and Hasson's results provide evidence that LLMs such as GPT-2 "capture subtle statistical dependencies reflecting syntactic, semantic, and pragmatic relationships among words" and hence seem to "support this shift from a rule-based symbolic framework to a vector-based neural code for processing natural language in the human brain."[6]

The second series, led by Evelina Fedorenko at MIT, makes somewhat similar comparisons and claims to have discovered "the language network," which differs from previous language-to-brain mappings.[7] Fedorenko's group has been studying the properties of this rather large brain volume, in quest of a neural theory of linguistic cognition. As we have seen before, looking at experiments through a magnifying glass is sometimes critical to their evaluation, so please bear with me, as I am quite convinced that taking a minute and look more closely at these two projects would pay off.

LLM–Language–Brain Relations Seem to Converge on Next Word Prediction

At the center of the Princeton project are measures taken from GPT-2, from behaving humans, and from the brains of epileptic patients undergoing a complex procedure similar to that undergone by neurosurgeon Edward Chang's patients (who helped to discover the neural substrate for abstract phonetic features; see chapter 7): surface electrodes are placed directly in clinically predetermined locations on patients' cortices, and as part of the clinical battery, their brains' electrical activity is recorded while they are performing a next word prediction task. The collected data enables a comparison between human behavioral responses, as well as their brains' and the LLM's. Similarity in performance would result in a high correlation between all three measures, and indicate that they predict the next word to the same degree.

A similar program, but with different test materials, was adopted by Fedorenko and her colleagues. Instead of an unconstrained recorded text, they devised a "localizer" task (inspired by an earlier fMRI test battery that has been said to quickly localize face recognition in each individual participant).[8] Like its face predecessor, this program aims to capture brain areas that support receptive language mechanisms and to set them apart from areas from which nonlinguistic perception and cognition emanate. Fedorenko's language task is said to be efficient and general, identifying in each participant a function that resides in roughly the same anatomical region. Participants hear or read meaningful sentences, and also controls, that are typically meaningless sequences of words or word-like syllable strings. This enables Fedorenko to zero in on a similar set of activated regions in a matter of minutes. The difference between test and control conditions, so goes the reasoning, isolates "sentence processing" in the brain. A second step studies the properties of these areas in detail, with the hope of testing specific models against the measurements that the "localizer" task has obtained.

Indeed, in a series of publications, Fedorenko and her team describe a test battery that they presented to healthy participants (and in some instances, to neurosurgical patients with electrodes installed on their cortex as I already described). Like the Princeton group, they expect to find correlations between the responses of GPT-2, the brain, and behaving humans.

Preliminary Critique: Methodological Requirements Are Not Met

Before getting into the actual results, let's pause to reflect on possible outcomes of these projects and what they might mean. Suppose that the language tasks probe all or at least most aspects of the broad range of our linguistic functions. Suppose further the best possible results are obtained and all the participants—the behaving humans, their active brains, and GPT-2—predict the next word perfectly, with perfect between-group correlations. Would such results lead to the conclusion that the machine is a scientific theory of language mechanisms in the human brain? Both groups of researchers seem to answer this question in the affirmative. And yet, even with such ideal (and imagined) results, Cao and Yamins's requirements for a theory are far from being met: excellent predictions and perfect correlations would nonetheless give no evidence that GPT-2 "correspond[s] to components, activities, properties, and organizational features of the target mechanism[s]" in the brain, nor would "the dependencies posited among these variables in the model correspond to causal relations among the components of the target mechanism." That is, even if all stars were to be properly aligned, we would have no analysis of the brain's or the model's *components* of language, although as we have already seen (in chapter 7), multiple results suggest that the linguistic brain is divided into grammatically structured pieces.

I believe that my final conclusion can already be anticipated: namely, that the way both groups probe the linguistic functions of their participants (healthy humans and neurosurgical patients, their brains, GPT-2) stops short of providing a realistic picture of their language capacities. Still, these studies are serious, they deserve careful scrutiny; moreover, I haven't told you what their findings are. This is where we go next.

The Princeton Studies

A pivotal element in these studies has obviously been the task that participants were carrying out. It was a recorded story, presented to the three groups (via headphones to the neurosurgical patients and in written form to the healthy humans and to GPT-2). The patients listened to the story passively, and the implanted electrodes recorded their brain activity as they were listening; the healthy participants viewed the same story on a screen, through a moving window, and performed a next word prediction task, with a fixed length moving window (that is, participants saw a fixed number of words from the text and were asked to predict the next one; then, they were shown a new word, and the earliest word they had seen vanished; and so on). GPT-2 was given a transcript of the story, likewise presented through a moving window; it returned predictions for each word. The researchers' analysis focused on the predictions, as measured at a time point just prior to a word's appearance.

Human participants made correct predictions at a 28% rate; GPT-2 performed better, and it naturally improved when given access to longer word strings before making its prediction, which lowered the entropy of incoming words. In about 10% of the electrodes placed on patients' brains, the levels of electrophysiological activity recorded during and immediately prior to the appearance of a word were correlated, as shown in figure 8.2. This correlation was interpreted as evidence for next word prediction. All three measures—brain, behavior, and GPT-2—were correlated.

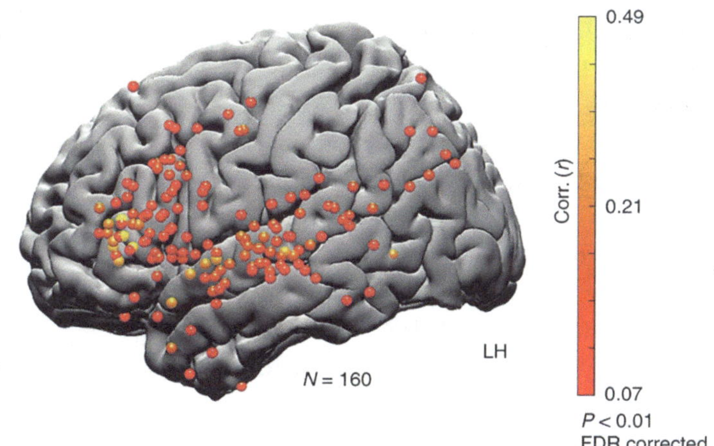

Figure 8.2
The position of 160 electrodes on the brain's surface (pooled from all 9 patients) that displayed a signal increase for the correct next word (or rather, its word embedding). The correlation goes from yellow (low) to red (high). Reproduced by permission from A. Goldstein, Z. Zada, E. Buchnik, M. Schain, A. Price, B. Aubrey, S. A. Nastase, et al., "Shared Computational Principles for Language Processing in Humans and Deep Language Models," *Nature Neuroscience* 25, no. 3 (2022): 369–380.

(continued)

The second study used the same data set to focus on three patients and provided a more detailed description of the neural code in which these words were expressed, as extracted from electrodes placed in Broca's area. It showed similarities between this code and the code produced by GPT-2 for these words. It also showed that signals extracted from adjacent brain areas are much less similar to GPT-2's and that relations between words that are expressed by a "symbolic" code are also less similar to what the brain produced. The study concluded that "deep language-model-induced representations of linguistic information are more aligned with brain embeddings sampled from [the inferior frontal gyrus] than symbolic representation."

In sum, these highly sophisticated studies found correlations between human behavior, machine, and brain activity, when responses were evoked by a text. If valid, these claims should lead to the rejection of grammatical representations as the correct description of the brain's linguistic code. Before scrutinizing these claims more carefully, let me move to the MIT set of studies whose authors draw similar conclusions.

Like the Princeton group, these MIT researchers sought a connection between the output of various LLMs, the activation pattern recorded from the collective functional region at issue, and associated behavioral patterns.[9] The correlation between the scores of GPT-2, brain activity, and behavior was strikingly high and, moreover, higher than that of any other LLM of the many that they tested. On the face of it, this is a remarkable set of results, which led them to claim that LLMs "are beginning to approximate the brain's mechanisms for processing language."

Let me pause to reflect on the broader significance of these claims: if true, they go way beyond the Turing test. These researchers claim not only that the model they have simulates human language behavior, as Turing had modestly hoped, but also that it provide the *true* neurological model of how the human brain brings about this simulation. That's a very tall order, which deserves careful scrutiny.

The MIT Studies

Fedorenko and colleagues contrasted meaningful sentences with lists of scrambled words and meaningless syllables. The brain areas that responded more strongly to the former were marked in each individual participant, then pooled together. They dubbed the resulting contrast map "the language network," seen in figure 8.3.[10]

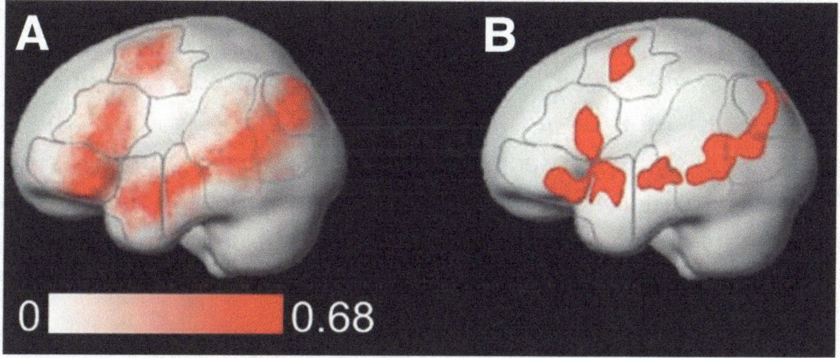

Figure 8.3
Scores of an fMRI experiment with healthy adult participants, who were given a task that involved reading meaningful sentences, and contrasted with the scores of their reading of meaningless sequences of pronounceable nonwords. Depicted is (A) a map that combined the scores of the whole group of participants (hence the fuzzy borders) and (B) the map of an individual participant. Reproduced by permission from I. Blank, N. Kanwisher, and E. Fedorenko, "A Functional Dissociation Between Language and Multiple-Demand Systems Revealed in Patterns of BOLD Signal Fluctuations," *Journal of Neurophysiology* 112, no. 5 (2014): 1105–1118.

The group map (A) shows participants whose scores converged on a very large area of the left hemisphere (and displays just 68% of the participants, who were at the maximal value, meaning that this group map excludes about one third of the participants that had been tested). The boundaries of this area are murky, because it is drawn on the basis of a large group of participants. Right-hemispheric parts of the alleged "language network" are not shown; an individual map (B) obviously has much sharper borders.

Next, Fedorenko and colleagues went looking for various cognitive properties of the marked area. For example, they attributed "formal linguistic competence" to their "language network," claiming that it "is sensitive to linguistic regularities" and arguing that this pattern of activation contrasts it with other brain networks serving other cognitive functions.[11]

The Main Critique: LLMs Are Not a Theory of the Linguistic Brain

Both groups have expressed optimism and excitement about LLMs as brain models for language. I wish I could be as positive. I contend that, contrary to their claims, the works discussed above provide no evidence that any current LLM models the brain's language processing device. In addition to the preliminary critique above, the main critique consists of several points. First, the neurolinguistic benchmarks by which these groups seek to sample performance are impoverished—the simple tests they deploy are plainly inadequate; second, LLMs may be good technologies, but they stop short of delivering a scientific theory of human language function; third, the studies in question display very poor anatomical precision, despite the existence of a rich neurolinguistic record, which points to a need to break language down into its component parts. This record is left unaccounted for. These are potentially devastating criticisms, so I'd better be a bit more detailed.

A. Neurolinguistic Benchmarks Are Far from Adequate

Both groups have been using language tests to probe human, machine, and brain activities. But what exactly are these tests, and what can we learn from the fact that they gave rise to the next word prediction behavior? We have already seen the critical importance of language test structure for the probing of human behavior and brain activity as well as machine behavior. Throughout this book we have seen both the necessity and usefulness of linguistically structured tests, leading to the conclusion that they are the ultimate evaluation tool for linguistic ability. In the context of the brain, we saw it through the remarkable success of Chang's neurosurgical group in identifying neuronal clusters that respond to abstract phonetic features. These researchers presented their participants with a highly structured corpus, which consisted of *the complete phonetic inventory of American English*. Without such a corpus, their conclusions would be unwarranted. As we have seen in previous chapters, scoping the broad range of the brain's sentence processing abilities is a truly formidable task. An informative "localizer" must include good representatives of this whole range of sentence probes (and their controls). Thus, testing the totality of neural processes that underlie sentence processing is a complex project that involves, at a bare minimum, a wide range of sentence types, chosen in a manner that is both theoretically and empirically reasoned: structural choices must be made; the relative proportion of each sentence type within the test needs to be properly evaluated; and test contrasts have to be constructed by control conditions, resulting in a well-structured, comprehensive test that isolates the cognitive component at issue and at the same time is short. We must therefore ask: did the above-discussed work on sentence processing in the brain employ the same high standard as the work on phonetics? Unfortunately, the answer is negative. The scope of the language materials is very narrow, and their structure stops short of delivering a proper test.

The Princeton group claims that its LLM-based method "can capture many aspects of the latent structure of human language, including syntactic trees, voice, coreferences, morphology and long-range semantic and pragmatic dependencies." But this claim is backed by no evidence, because syntactic trees, voice, coreferences, morphology, and long-range semantic and pragmatic dependencies are not tested in any of their studies, whether directly or indirectly; nor is there an indication as to how such relations could be tested in their paradigm.

Instead of following the path I charted above, the Princeton test amounts to an auditory presentation of a famous segment of National Public Radio's popular *This American Life* called "Monkey in the Middle," which tells an amusing story of a monkey who pushed the button of a photographer's field camera and took a selfie that ended up being widely distributed and resulted in (rather juicy) legal disputes. Curious as this story may be, its inventory of utterances does not comprise a representative sample of most, let alone all, aspects of language. It is, rather, an unselected text that mostly consists of active declarative sentences: *Our story begins deep in the rainforests of Indonesia on an island called Sulawesi. A few years ago, the photographer David Slater traveled there from his home in England to photograph a troop of monkeys.*

The MIT researchers claim that "participants read semantically and syntactically diverse sentences," contrasted with meaningless sequences of words or syllables of roughly equal length. In some cases, they were given "naturalistic" texts—stories found on the web. A review of the sentence materials, however, indicates that the variety of sentences used was extremely limited, both syntactically and semantically, as it mostly consised of simple active declarative sentences (with an occasional embedding). For example, *Senator Hubert Humphrey is obviously a man with a soul and heart*, contrasted with meaningless sequences such as *quey etoa pimning strosh chsture en freir polyll denumpments ro veep urper*. The stories used are short and simple, and the controls are nonsensical sequences of syllables. A simple design like this is an advantage, at times,

(continued)

but not here: if one seeks to study the manner by which the brain processes language in all its complexities at a level of sophistication that is similar to that of vision neuroscience, then the bar must be higher: for these standards to be met, simple and complex syntactic types must be used, in a manner already discussed.

Both groups' language materials, which aimed to evoke behavioral, neural, and machine responses, actually consisted of arbitrarily selected sentences. As such, the nature of their language localizer is not comparable to the theoretically motivated selection of materials that is customary in vision, audition, or phonetics. No wonder, then, that in the Princeton studies, a random collection of electrodes throughout the brain was modulated by the input and that, in the MIT series of studies, the brain area activated was huge and bihemispheric, localized in a set of regions whose borders converge on no known anatomical landmarks.

The Princeton studies concluded that, compared to network code, "symbolic" representations produced weak brain–machine correlations. But upon careful examination, their "symbolic" representations contain no syntactic or semantic rules but are, rather, representations of arbitrarily assigned categories, developed and used for automatic word tagging. These do not contain rules, nor are they stated in a format that is part of any known linguistic theory. They may eventually turn out to be the correct ones. However, at this point, basing an analysis on this classification method cannot lead to a conclusion regarding the advantage (or lack thereof) of one system of representation to another.

The unsystematic sentence samples modulate bihemispheric clusters that are not only too large, but also . . . too small: the test materials lack examples of many syntactic and semantic components that are essential for human communication. For example, they hardly contain any questions, and they do not properly represent imperatives, quantifiers, modal predicates, embeddings, displacements, ellipses, and a host of intrasentential dependency relations. Nor do they evince a systematic choice of phonetic and phonological materials. *This set of tests stops short of selectively localizing any recognizable cognitive or linguistic component, let alone a syntactic one.*

The near uniformity of sentence types in these tests almost amounts to multiple presentation of the same sentence, from a grammatical perspective.

Imagine that you hear the same sentence structure one million times (*John loves Mary*; *The man saw the woman*; *Elephants eat grass*; etc.). Before long, your brain would get adapted to this Subject–Verb–Object sequence, and your behavior would align its expectations to it, as would your brain, which would likely make behavioral and brain responses different than if you heard a multiplicity of sentence types. Recall that the goal of both research groups has been to investigate sentence processing (the Princeton group even claimed that its model has the capacity to capture specific complex syntactic relations). What was actually tested turns out to be an extremely narrow range of structures that provides no empirical basis for such claims. Some readers may recall the movie *The Shining*, in which the protagonist, Jack Torrance, seems to be writing a novel, which turns out to consist of repetitions of the same sentence: *All work and no play makes Jack a dull boy*. Linguistically, the tests in both the Princeton and MIT studies bear some similarity to Jack Torrance's novel: these researchers believe that they are testing language, while repeatedly testing structurally similar sentences.

B. LLMs Do Not Deliver a Theory of Human Language Function

My second criticism starts by questioning the suitability of GPT-2 as a model of human language, independent of brain implementation. Here, the picture is clear: based on the conceptual, methodological, and empirical arguments I gave in chapters 5 and 6, no LLM thus far created can qualify as a theory of human language—these models are unconstrained and have proven unsuccessful in correctly classifying complex cases as ±grammatical.[12] We therefore *know* that GPT-2 is not a candidate theory. Digging deeper, we see that the specialized attention heads in GPT-2 classify the grammaticality of sentences with MOVE at a mediocre level (under 80% accuracy).[13] These models thus do not qualify as scientific theories of brain function, at least not in their current form.

C. Brain Localization of Processing Components Is Far from Precise

My final point regards brain localization. Our brain is noisy, and language processing is a highly complex activity. Listening to the linguistic brain and deciphering its signal is, moreover, difficult. Still, when one compares past work on the localization of language functions to that of the Princeton and MIT studies, the anatomical resolution of the latter is low (as seen in figures 8.2 and 8.3), certainly when compared to the sharper picture obtained by experiments that use more focused language materials: chapter 7 displayed relatively sharp images of small brain locations for specific components of language, aligned with microscopic anatomy, thereby opening the door to a detailed functional anatomy of a mosaic of syntactic and semantic processes. Most of these results, displayed at a much higher functional as well as anatomical resolution, are not discussed by either set of studies. This is no accident: for all we know, the fine linguistic distinctions that have been found in the brain cannot be accommodated by GPT-2 or by any of its successors.

Coda: Can LLMs Model the Brain's Language Mechanisms?

Much is required for an apt characterization of receptive language, at the empirical, methodological, anatomical, and linguistic levels. Chapter 7 concluded with a divide-and-conquer (or better yet, parcellate-and-model) agenda for neurolinguistics, which would begin with the parcellation of the microanatomy on the one hand and the function on the other hand. Overall modeling, such as that done by the direct implementation of LLMs, misses the point: not only does it fail to divide language into its components (as is necessary for both theoretical and empirical reasons), but also, it provides no tools for the discovery of anatomically identifiable pieces of cortical tissue that are entrusted with specific language functions. The prospects of such a global analysis are, therefore, rather grim. More focused projects that seek to model specific parts of language in the brain are more likely to succeed, similar to those on the visual and sensorimotor systems.

Taking Stock

This chapter was the second to get under LLMs' hood. We saw:

- Studies of LLM, human, and brain performance and the relation between their results
- That in the visual domain, methodological principles are well-articulated and models are constructed in keeping with what is known from brain neurophysiology and microanatomy
- That in language studies this is not always the case
- That all reviewed neurolinguistic studies that incorporate LLMs have used tests that are only weakly informative, mostly due to poor selection of localizer tasks and test materials
- That the conclusion that the brain's specialized underlying mechanisms are LLMs is premature, to say the least
- That the claim that a novel brain location of language mechanisms has been identified is largely unwarranted
- That a more viable strategy for the discovery and modeling of the linguistic brain is (a) parcellating both the function, using linguistic tools and the structure, using anatomical tools, followed by (b) deep network modeling of the parts

Epilogue: Shall We Work Together?

"Other maps are such shapes, with their islands and capes!
But we've our brave Captain to thank:"
(So the crew would protest) "that he's bought us the best—
A perfect and absolute blank!"

This was charming, no doubt; but they shortly found out
That the Captain they trusted so well
Had only one notion for crossing the ocean
And that was to tingle his bell.
—Lewis Carroll, *The Hunting of the Snark*, 1876

I may not have gone where I intended to go, but I think I have ended up where I intended to be.
—Douglas Adams

An AI Fantasy?

This book's subtitle is *Chomsky, Our Brain, and the AI Fantasy*. We talked about linguistic theory, inspired by Chomsky; about our linguistic brain; and about AI in its current incarnation—but no fantasy has been mentioned, at least not explicitly.

What's a fantasy? ChatGPT's answer (apparently based on *Merriam-Webster*): "Something imagined or created in the mind, typically involving elements that are impossible or unlikely to occur in reality." Do the arguments in this book support the last, somewhat provocative, part of its subtitle? I believe they do. The fantasy I described is the thought that LLMs alone can provide a true model of human speakers' linguistic abilities and brain activity, which I believe won't happen. For this ambition to come true, LLMs would need to pass a much higher bar than the one set by Alan Turing through his test: not only would they need to mimic human language behavior successfully, but also, they would need to constitute *the* true model of the neural networks that underlies linguistic capacity. I have given you good reasons to believe that without incorporating linguistic theory, this is pure fantasy. Let me quickly summarize the arguments, in text and in a table.

The arguments were of three types. *Empirically*, I provided multiple facts—linguistic, anatomical, and neurolinguistic—that suggest that the proposed models, while creative and interesting, must incorporate linguistic theory, or else they remain off the mark as they do not fit the rich available experimental record (which they simply ignore, for the most part). *Conceptually*, I have argued that these models are unconstrained and are moreover opaque; that is, on their own, they are too large to be interpretable. A modular design that responds to grammatical and neurological needs will likely enhance our understanding, without which, we have no good story to tell. Finally, from an *engineering* point of view, I hinted that these models (whose authors keep citing Moore's Law concerning the growth rate of computational power) seem overly muscular. The rarity of engineering efforts to constrain them threatens to lead to energy-hungry machines that kill flies with cannonballs. As theories, none of the current ones is parsimonious. Luckily, neurolinguistics has an alternative scientific agenda, which I presented briefly in chapter 7.

Here are some brief bullet points of the story as it unfolded:

- TGG delivers a scientific theory that describes the knowledge humans possess about their language and that is consistent with neurolinguistic results. The implementability (and runnability) of this class of theories is far from impressive, and its proponents may have been negligent is seeking dialogue with the neural network community. However, we currently have an impressive class of linguistic theories that are transparent and truly explanatory.
- LLMs are task-oriented language technologies, having produced impressive simulations of a speaker/hearer that are runnable on a computer. However, they fail on a surprisingly broad range of psycho- and neurolinguistic tests, they are nontransparent, and they do not deliver a biologically or conceptually feasible (let alone constrained) scientific explanation of the human linguistic ability and its neural underpinnings.

- Neuroanatomy and physiology have provided detailed perspectives about the underlying biology of language and speech. A rich record indicates that the linguistic brain, while massively interconnected, is also highly modular, in a manner that is sensitive to linguistic generalizations. Anatomical linguistic modules, or "brain areas that compute linguistic functions," seem to be structured in terms of the microscopic properties of the neural tissue that supports them.
- Bridging the gap between anatomical structure and cognitive function is now almost routinely done in vision neuroscience but is still rare in neurolinguistics. This can and should change.

I have addressed a variety of issues, on which the two approaches agree and disagree. The main disagreements, advantages, and deficiencies are summarized in table E.1.

The Necessity of Working Together

Language, it appears, is very deeply human, presenting humanity with a vast array of problems and puzzles. This leads to a clear conclusion: *we must work together!* Linguists should indicate where to look, how to construct benchmarks, and how to structure the functions of our model; neurolinguists should tell us how to modularize our network and how to probe humans and brains; and network experts should build the networks and provide mechanistic explanations for the linguistic generalizations. What could be more natural than this collaborative effort?

Without linguists, claims about the machines' performance often end up being embarrassingly unfounded; without network specialists, they are not even built; and without neurolinguists, no coherent perspective about brain function can be developed.

Before We Say Goodbye: A Learning Success and a Learning Failure

If your brain analyzed a sentence successfully, then your next encounter with the same type of sentence (e.g., a question, an active declarative sentence, a conditional sentence) would be faster and less taxing, even if the words were different. This has been demonstrated time and again, and it has been dubbed *syntactic priming*.[1] It amounts to a demonstration that

Table E.1

Issue	LLMs	TGG
1. **Innateness.** Does language have an innate knowledge component?	**NO**	**YES**
2. **Child development.** Does the system solely depend on input to get to its adult steady state? What are the learning methods?	**YES** Input-dependent, sensitive to quantitative aspects of the input	**NO** Little sensitivity to quantitative aspects of the input
3. **Competence versus performance.** Should the theory make this distinction?	**NO**	**YES**
4. **Constraints.** Is the model or theory constrained, or can it grow indefinitely?	**NOT** really. The few constraints one finds are imposed by hardware limitations	**YES**, there are many constraints, built into the theory, enhancing explanatory adequacy
5. **Neural description.** What is the right level of description? Does individual variation matter?	Neural networks/connectivity/layers/weights, simulations	Functional anatomy at the level of microscopic anatomical parcellation; causally modeled connectivity
6. **Brain localizers.** What method is used to parcellate the brain into its functional parts?	Electrodes; functional ROIs (regions of interest) based on gross distinctions	Anatomic ROIs, based on fine anatomy, mapping with cytoarchitecture
7. **Performance tests.** What is the right benchmark? How to build a general evaluation method for the system?	Turing test: "imitation game," "creativity"; gross localizers; behavior/response correlations in next word prediction tasks	Comprehension, judgment, and verification tasks on fine theoretically motivated linguistic contrasts
8. **Language evaluation.** What are the inventories/test materials?	Language "in the wild," structured tests (BLiMP, CoLA); lab-specific localizers	Multiple theory-driven choices
9. **Runnability.** Can the theory be run on a machine?	**YES**	**NO**
10. **Transparency.** Can a failure be attributed to a specific theoretical construct or component?	**NO**	**YES**
11. **Interpretability.** Do we understand why the system behaves as it does?	**NO**	**YES**

your brain learned a generalization—that two sentences that are built out of different words but that share abstract structure are similar.[2]

LLMs do not analyze syntactic structure, but they are said to be fast learners. To their proponents, their ability to generalize from examples indicates amazing learning abilities.[3] So here is a new linguistic way to assess this ability, whose existence underscores the need for collaboration between our communities: if the LLM is a fast learner, then once it is taught the properties of a particular structural relation, it should easily generalize to new examples. Pursuing this logic, I set up a two step test of ambiguity detection learning—a task that humans can perform easily—and conducted it with ChatGPT-4o.

Step one tested the machine's ability to detect a basic structural ambiguity discussed in chapter 1. I presented it with an ambiguous sentence: *The first police commissioner was a woman.* (Structure$_1$: *the [first [police commissioner]].* Meaning$_1$: the first commissioner of some police force in the common ground was a woman. Structure$_2$: *the [[first police] commissioner].* Meaning$_2$: the commissioner of the first police force in the common ground was a woman.) Instead of asking whether the sentence was ambiguous, I requested that the machine paraphrase it; the machine returned a single unambiguous sentence, the same as the original sentence but with two words substituted: *The <u>initial</u> police commissioner was <u>female</u>.* (Later, I tried other example sentences, and at times, it was able to identify the ambiguity, although paraphrasing was more difficult.) I replied that my first sentence was ambiguous; ChatGPT agreed to be corrected (as it almost always does, apparently hand-programmed to refrain from challenging its interlocutor), "admitting" that it had previously failed to detect the ambiguity; after minimal prompting, it successfully expressed both meanings. Learning seemed to have occurred.

The critical part was step two, in which I tested the machine's ability to generalize. If the lesson had been learned, I reasoned, and if the machine can generalize, then it would quickly identify the ambiguity in a new sentence that has the same structural properties. This is a true test of generalization ability and a task easy enough for any reasonable human. I therefore asked the machine to paraphrase a close relative of the previous sentence: *The first army commander was six foot tall.* Most humans would succeed in detecting the ambiguity here, especially if asked in the immediate context of the previous instance of structural ambiguity. This is generalization. To my surprise, I must admit, ChatGPT-4o failed to generalize. Learning had not taken place.

I encourage you to play and invent such games, and I hope you will be amused. More importantly, I sincerely hope that a more serious lesson is learned here: such failures can only be detected once linguistic considerations are invoked (and these are rather minimal). And yet, these considerations are never invoked in the discussion of LLMs' language learning. This type of two step test can be cooked up by anyone with a fairly basic background in linguistics. As LLMs are not equipped with a method to generate predictions and tests of the type described, there is no way to discover these machines' limitations (or virtues, for that matter). It is rather shocking that after a decade of research on learning in LLMs and after much talk about "scientific revolutions" and "paradigm shifts," such simple tests have not been conducted. May it serve as yet another argument for the acute necessity of a collaborative effort between the disciplines.

What to Do: A Sketch of a Recipe

In the preceding chapters, I sketched an ongoing research program that aims at discovering a mosaic of microfunctions and their microanatomical correlates. I gave some hints that suggest the validity of this approach, and I pointed to similarities between it and leading approaches to the study of brain mechanisms for vision. Pursuing this line of reasoning would not only be collaborative in a manner that would cross disciplines and let groups with varied interests teach each other what they know about the field, it would also shift the direction of research from a study of generic networks to investigations of neurologically supported micromodular local networks. Thus, starting off with modules and constraints on their operation, studying the cytoarchitecture of the supporting tissue, and using all this for the construction of small networks is to my mind the right way to achieve progress working together. Even engineers would gain from it, as linguistic knowledge would act as a power saver, helping them trim redundant processes. This is not a fantasy: it's a concrete research program, which would lead to major progress in the study of neurolinguistics and neurocognition.

Noam Chomsky repeatedly said that "linguistic theory is part of psychology, ultimately biology." Adopting the strategy I have sketched would hopefully bring this research program closer to reality.

An Afterthought: AI and MI, October 8, 2023

The Human-Machine team enables a new potential to use AI to build a "smart area" for protecting borders. The ability to connect various bits around the border through machine-learning can help control these areas and protect the borders.

—Brigadier General Y. S. (Yossi Sariel), Commander, Unit 8200, Israel Defense Forces, *The Human-Machine Team*, 2021

On October 8, 2023, when it became clear that the line with Gaza had collapsed, I knew how this book would end. The Gaza Strip, spanning altogether 141 square miles (population: 2.1 million), has for years been thoroughly monitored by eavesdropping devices, camera-equipped drones and balloons, and much else. The Israeli intelligence community, especially the Military Intelligence (MI) section of the Israel Defense Forces, was said to have divided the Strip into data-generating tiny squares, fed into artificial intelligence (AI)–guided devices. Signals intelligence technologies were developed by the famed Unit 8200, which has long attracted the best and the brightest young Israeli minds, becoming the pride of the tech nation and the largest Israel Defense Forces unit. Scores of this unit's graduates (some of whom would become my students) were hired at top positions in the AI industry worldwide; its commander preached a "Border with 'AI Walls'";[4] and its top brass bragged that "our computers can indicate whether the information is valuable, as well as its value level."[5]

Every Israeli knew that the borders were watched and secured; that Unit 8200 was there, tirelessly analyzing instreaming big data. The press assured us that everything was fed into superfast computers and chewed up by LLMs. We were also told that, aware of past failures, the MI continually consulted with smart devil's advocates, who kept questioning basic assumptions and left no stone unturned.

Everyone adored them and slept well, and then, boom! On Saturday, October 7, 2023, this whole creation goes down the drain in less than a day: a murderous attack takes the start-up nation by complete surprise. Hamas staged a well-planned attack, felled the border fences in minutes, and its men murdered, mutilated, and violated some 1,200 innocent civilians, took hostages, and burned their homes. In retaliation, Israel declared a war on Hamas that has since killed dozens of thousands of Gazans, among them almost 20,000 children, wounded even more and has left huge numbers hungry, wasted, destitute, homeless, and displaced, with no end, solution, or hope in sight.

How was Hamas able to cook up such a surprise? Where were the MI and its Unit 8200? How could the MI's allegedly supreme capacity to eavesdrop and its unmatched AI abilities to predict the enemy's next step miss a huge plan in which thousands were involved?

Uri Bar-Joseph, the renowned military historian of the surprise Arab attack on Israel in 1973,[6] gloomily described in November 2023:

> the tendency in the Military Intelligence (MI) to take lessons of the past lightly, and rely almost exclusively on technological alerts. . . . I saw this tendency myself less than two months ago. In an event marking fifty years for the intelligence failure in . . . 1973, I presented to MI's senior officers my explanation for the roots of this failure. I said that first and foremost, it was the personalities of some past senior officers who stuck to the prevailing conception to the very last minute, despite the fact that all the information in front of them was screaming "war!" A subsequent lecture at that event was about an experiment, in which available pre-war data were fed to an AI machine, in order to examine whether AI can substitute human thinking. The discussion period was mostly focused on the machine's abilities to identify threats. The psychology of evaluation failures was not of interest to the MI evaluators. As I left the discussion, it was clear to me that the lessons from 1973 were not learned. Not for a moment did I imagine that it would become so painfully and shamefully clear at this speed.[7]

By early December, it was becoming clear that heaps of evidence had been sitting right at the MI's door. Hard questions about basic assumptions were not asked, as the burden was shifted to LLMs, which did not converge on the right evaluations. On-site human observers continuously sent clear warnings, but neither machine nor high command listened.[8] What had been touted as the greatest military intelligence service ever fell victim to a gigantic deceit by Hamas's guerrilla army, in the face of heaps of evidence for an imminent attack. All warnings went unnoticed.[9]

Strategists and experts will spend years discussing this tragic surprise, but one thing already seems clear (in fact, almost a consensus among experts): the MI and its satellite organizations became addicted to high tech, relied on developers, and failed to think hard about the right characterization of the problems that these technologies were expected to solve. They believed that their devices were so powerful that they could offer a solution to anything and spent insufficient amounts of thought on whether they were choosing the right problems to solve. Thus, seeking prediction without clear definitions of the problems that lie ahead, ignoring large datasets and relevant biases in the input data, and failing to evaluate correctly or at least consider alternatives seem to have been central factors in Israel's worst intelligence failure ever.

Why am I dwelling on intelligence at war in a book on language? Because this (still ongoing) Middle Eastern tragedy is highly relevant to the debate on the human linguistic ability, to which this book is dedicated. A neurolinguist living in Tel Aviv, I find it difficult to miss the similarities between the atmosphere within the Israeli MI prior to October 7 and certain prevailing sentiments in the AI community that models natural language and its neural bases.

I love technology, and I love science—incontrovertibly among the best products of human culture. But overreliance on technology (and to some extent on science) may lead to complacency. Hence the fate of the AI wall story, a fantasy that everyone loved to believe, until it collapsed, as border fences fell in a matter of minutes. Tragically, the largest and smartest unit of one of the strongest military organizations on the globe failed to predict a surprise ground attack by guerrillas riding in low tech white Toyota Hilux pickups.

In addition to complacency, overreliance on technology threatens to lead science astray. On the face of it, the practices of LLMs' proponents who build language technologies are strikingly similar to those of the Israeli MI. As Rich Sutton puts it, more technology, more big data, more memory, and more computation would make unsolved problems go away fast. This belief has become widespread. Indeed, when my AI-heavy neuroscientist colleagues are presented with problems that ChatGPT has had (something I have been showing them for a while), they respond, unwaveringly, "if ChatGPT would only train properly, he/she/it will eventually get it." Their belief in the power of AI appears to run deep.

And yet, a mere year and nine months ago (we are now in July 2025), a critique of the Israeli MI and its AI capacities would have been dismissed as heresy. The Israeli technological disaster, I believe, should echo far and wide. This book, aimed at both lay readers and experts, attempts to spell out problems with the use of language technologies in science and elsewhere. In it, I have tried to consider alternatives, in the hope of learning an important lesson for a better future. Presently, I approach stories about great technologies and, to some extent, about science with awe but also with suspicion and fear. And in the case dearest to my heart—human language—I see great successes but also serious problems and perils, and I wish that readers would keep this in mind.

Notes

Prologue

1. Press release, The Nobel Prize in Physics 2024, The Royal Swedish Academy of Sciences, October 8, 2024, https://www.nobelprize.org/uploads/2024/11/press-physicsprize 2024-2.pdf.

2. https://www.youtube.com/watch?v=-icD_KmvnnM.

3. E. M. Bender, T. Gebru, A. McMillan-Major, and S. Shmitchell, "On the Dangers of Stochastic Parrots: Can Language Models Be Too Big? 🦜," in *Facct '21: Proceedings of the 2021 ACM Conference on Fairness, Accountability, and Transparency* (Association for Computing Machinery, 2021), 610–623; "Do You Need a New iPhone?+Yuval Noah Harari's AI Fears+Hard Fork Crimes Division," *Hard Fork* (*New York Times* podcast, September 13, 2024.

4. G. F. Marcus, *Taming Silicon Valley: How We Can Ensure That AI Works for Us* (MIT Press, 2024).

5. Comment during panel discussion "Research on Intelligence in the Age of AI," Center for Brains, Minds, and Machines (October 7, 2023; clip posted on *Statistical Machine Learning* YouTube channel, https://www.youtube.com/watch?v=urBFz6 -gHGY).

6. For example, "The False Promise of ChatGPT," *New York Times*, March 8, 2023.

7. *Taming Silicon Valley: How We Can Ensure That AI Works for Us* (MIT Press, 2024).

Chapter 1

1. V. Vergiani, "Pāṇini's Aṣṭādhyāyī: A Turning Point in Indian Intellectual History," *Rivista degli Studi Orientali* 42, nos. 3–4 (2019): 11–35.

2. P. Kiparsky, "Pāṇinian Linguistics," in *Concise History of the Language Sciences*, ed. E. F. K. Koerner and R. E. Asher (Pergamon, 1995), 59–65.

3. N. Vidro, "Grammars of Classical Arabic in Judaeo-Arabic: An Overview," *Intellectual History of the Islamicate World* 8, nos. 2–3 (2020): 284–305; A. Arnauld and C. Lancelot, *General and Rational Grammar: The Port-Royal Grammar*, trans. J. Rieux and B. E. Rollin (Mouton, 1975).

4. L. Bloomfield, *Language* (New York: Henry Holt, 1933).

5. Z. S. Harris, *Methods in Structural Linguistics* (University of Chicago Press, 1951).

6. B. F. Skinner, *Science and Human Behavior* (Simon and Schuster, 1965).

7. B. F. Skinner, *Verbal Behavior* (Appleton-Century-Crofts, 1957), https://doi.org/10.1037/11256-000.

8. C. E. Shannon, "A Mathematical Theory of Communication," *Bell System Technical Journal* 27, no. 3 (1948): 379–423.

9. G. A. Miller, *Language and Communication* (New York: McGraw-Hill, 1951).

10. E. Sapir, *Language: An Introduction to the Study of Speech* (New York: Harcourt, Brace, 1921); B. Whorf, *Language, Thought, and Reality: Selected Writings of Benjamin Lee Whorf*, ed. J. B. Carroll (MIT Press, 1956).

11. N. Chomsky, "Three Models for the Description of Language," *IRE Transactions on Information Theory* 2, no. 3 (1956): 113–124.

12. M. D. Hauser, N. Chomsky, and W. T. Fitch, "The Faculty of Language: What Is It, Who Has It, and How Did It Evolve?" *Science* 298, no. 5598 (2002): 1569–1579.

13. (The Hague: Mouton, 1957).

14. M. Dalrymple, *Lexical Functional Grammar* (Brill, 2001); M. Steedman, "Categorial Grammar," in *The Routledge Handbook of Syntax*, ed. A. Carnie, Y. Sato, and D. Siddiqi (Routledge, 2014), 670–701; N. Hornstein, "The Minimalist Program After 25 Years," *Annual Review of Linguistics* 4, no. 1 (2018): 49–65.

15. J. Berko, "The Child's Learning of English Morphology," *Word* 14, nos. 2–3 (1958): 150–177.

16. P. Lipton, "Inference to the Best Explanation," in *A Companion to the Philosophy of Science*, ed. W. H. Newton-Smith (Blackwell, 2000), 184–193.

17. J. Searle, "Chomsky's Revolution in Linguistics," *New York Review of Books*, June 29, 1972.

18. N. Chomsky and G. A. Miller, "Introduction to the Formal Analysis of Natural Languages," in *Handbook of Mathematical Psychology*, ed. R. Luce, R. Bush, and E. Galanter (New York: Wiley, 1963).

19. N. Chomsky, *Aspects of the Theory of Syntax* (Cambridge, MA: MIT Press, 1965).

Chapter 2

1. J. A. Fodor, *The Language of Thought* (Cambridge, MA: Harvard University Press, 1975); *LOT2: The Language of Thought Revisited* (Oxford: Clarendon Press, 2008).

2. S. J. Gould and E. S. Vrba, "Exaptation—A Missing Term in the Science of Form," *Paleobiology* 8, no. 1 (1982): 4–15.

3. N. Chomsky, *Current Issues in Linguistic Theory* (Mouton, 1964); J. R. Ross, "Constraints on Variables in Syntax" (PhD dissertation, Massachusetts Institute of Technology, 1967).

4. N. Chomsky, "Three Factors in Language Design," *Linguistic Inquiry* 36 (2005): 1–22.

5. M. Steedman and J. Baldridge, "Combinatory Categorial Grammar," in *Non-Transformational Syntax: Formal and Explicit Models of Grammar*, ed. R. D. Borsley and K. Börjars (Wiley-Blackwell, 2011), 181–224; J. Y. Findlay, "LFG and Tree-Adjoining Grammar," in *Handbook of Lexical Functional Grammar*, ed. M. Dalrymple (Language Science Press, 2024), 2069–2125.

6. L. Rizzi, *Relativized Minimality* (Cambridge, MA: MIT Press, 1990).

7. J. van Craenenbroeck and T. Temmerman, eds., *The Oxford Handbook of Ellipsis* (Oxford University Press, 2019).

Chapter 3

1. "The Nobel Prize in Physiology or Medicine 1906," *NobelPrize.org* (n.d.), https://www.nobelprize.org/prizes/medicine/1906/summary/.

2. W. S. McCulloch and W. Pitts, "A Logical Calculus of the Ideas Immanent in Nervous Activity," *Bulletin of Mathematical Biophysics* 5 (1943): 115–133.

3. F. Rosenblatt, "The Perceptron: A Probabilistic Model for Information Storage and Organization in the Brain," *Psychological Review* 65, no. 6 (1958): 386.

4. Y. Bar-Hillel, "The Present Status of Automatic Translation of Languages," *Advances in Computers* 1 (1960): 91–163.

5. M. Minsky and S. A. Papert, *Perceptrons* (Cambridge, MA: MIT Press, 1969), 318–362.

6. D. O. Hebb, *The Organization of Behavior: A Neuropsychological Theory* (Wiley, 1949).

7. D. E. Rumelhart, G. E. Hinton, and R. J. Williams, "Learning Representations by Back-Propagating Errors," *Nature* 323, no. 6088 (1986): 533–536.

8. Y. LeCun, Y. Bengio, and G. Hinton, "Deep Learning," *Nature* 521, no. 7553 (2015): 436–444.

9. L. R. Gleitman, M. Y. Liberman, C. A. McLemore, and B. H. Partee, "The Impossibility of Language Acquisition (and How They Do It)," *Annual Review of Linguistics* 5, no. 1 (2019): 1–24.

10. E. G. Wilcox, R. Futrell, and R. Levy, "Using Computational Models to Test Syntactic Learnability," *Linguistic Inquiry* 55, no. 4 (2024): 805–848; N. Lan, E. Chemla, and R. Katzir, "Large Language Models and the Argument from the Poverty of the Stimulus," *Linguistic Inquiry* (advance online publication, August 30, 2024).

11. T. P. Lillicrap, A. Santoro, L. Marris, C. J. Akerman, and G. Hinton, "Backpropagation and the Brain," *Nature Reviews Neuroscience* 21, no. 6 (2020): 335–346.

12. G. Hinton, "Will Digital Intelligence Replace Biological Intelligence?" (Arthur Miller 2023–2024 Lecture on Science and Ethics, Massachusetts Institute of Technology, December 11, 2023), https://www.youtube.com/watch?v=iWPo7Yhg7Vc&t=2149s, 14:01.

13. A. Krizhevsky, I. Sutskever, and G. E. Hinton, "ImageNet Classification with Deep Convolutional Neural Networks," in *Advances in Neural Information Processing Systems 25 (NIPS 2012)*, ed. F. Pereira, C. J. Burges, L. Bottou, K. Q. Weinberger (Neural Information Processing Systems Foundation, 2012), https://papers.nips.cc/paper_files/paper/2012.

14. A. William, M. Miceli, and T. Gebru, "The Exploited Labor Behind Artificial Intelligence," *Noēma*, October 13, 2022.

Chapter 4

1. M. Meyers, G. Weiss, and G. Spanakis, "Fake News Detection on Twitter Using Propagation Structures," in *Disinformation in Open Online Media: Second Multidisciplinary International Symposium, MISDOOM 2020*, ed. M. van Duijn, M. Preuss, V. Spaiser, F. Takes, and S. Verberne (Springer, 2020), 138–158.

2. A. A. Markov, "An Example of Statistical Investigation of the Text *Eugene Onegin* Concerning the Connection of Samples in Chains," trans. G. Custance and D. Link, *Science in Context* 19, no. 4 (2006): 591–600 (originally published 1913).

3. S. T. Piantadosi, "Zipf's Word Frequency Law in Natural Language: A Critical Review and Future Directions," *Psychonomic Bulletin and Review* 21 (2014): 1112–1130.

4. C. E. Shannon, "A Mathematical Theory of Communication," *Bell System Technical Journal* 27, no. 3 (1948): 379–423.

5. G. Pardelli, M. Sassi, and S. Goggi, "From Weaver to the ALPAC Report," in *Proceedings of the Fourth International Conference on Language Resources and Evaluation (LREC'04)*, ed. M. T. Lino, M. F. Xavier, F. Ferreira, R. Costa, and R. Silva (European Language Resources Association, 2004).

6. A. Taylor, M. Marcus, and B. Santorini, "The Penn Treebank: An Overview," in *Treebanks: Building and Using Parsed Corpora*, ed. A. Abeillé (Dordrecht: Springer, 2003).

7. J. R. Firth, "A Synopsis of Linguistic Theory, 1930–1955," in *Studies in Linguistic Analysis*, ed. J. R. Firth (Oxford: Blackwell, 1957).

8. Edited by G. E. M. Anscombe (New York: Wiley-Blackwell, 1953).

9. Y. Lakretz, T. Desbordes, J. R. King, B. Crabbé, M. Oquab, and S. Dehaene, "Can RNNs Learn Recursive Nested Subject-Verb Agreements?" (preprint, January 6, 2021), https://doi.org/10.48550/arXiv.2101.02258.

10. J. Russin, S. W. McGrath, D. J. Williams, and L. Elber-Dorozko, "From Frege to ChatGPT: Compositionality in Language, Cognition, and Deep Neural Networks" (preprint, May 24, 2024), https://doi.org/10.48550/arXiv.2405.15164.

11. A. Vaswani, N. Shazeer, N. Parmar, J. Uszkoreit, L. Jones, A. N. Gomez, Ł. Kaiser, and I. Polosukhin, "Attention Is All You Need," in *Advances in Neural Information Processing Systems 30 (NIPS 2017)*, ed. I. Guyon, U. Von Luxburg, S. Bengio, H. Wallach, R. Fergus, S. Vishwanathan, and R. Garnett (Neural Information Processing Systems Foundation, 2017), https://papers.nips.cc/paper_files/paper/2017.

12. J. Devlin, M. W. Chang, K. Lee, and K. Toutanova, "BERT: Pre-Training of Deep Bidirectional Transformers for Language Understanding" (preprint, October 11, 2018), https://doi.org/10.48550/arXiv.1810.04805.

13. W. Vanderbauwhede, "Frugal Computing—On the Need for Low-Carbon and Sustainable Computing and the Path Towards Zero-Carbon Computing" (preprint, March 12, 2023), https://doi.org/10.48550/arXiv.2303.06642; Kudithipudi, D., Schuman, C., Vineyard, C.M., Pandit, T., Merkel, C., Kubendran, R., Aimone, J.B., Orchard, G., Mayr, C., Benosman, R. and Hays, J., 2025. Neuromorphic computing at scale. *Nature, 637*(8047), 801–812.

14. C. Singh, J. P. Inala, M. Galley, R. Caruana, and J. Gao, "Rethinking Interpretability in the Era of Large Language Models" (preprint, January 30, 2024), https://doi.org/10.48550/arXiv.2402.01761.

Chapter 5

1. H. L. Dreyfus, *What Computers Can't Do* (New York: Harper and Row, 1972).

2. T. J. Sejnowski, "Large Language Models and the Reverse Turing Test," *Neural Computation* 35, no. 3 (2023): 309–342, https://doi.org/10.1162/neco_a_01563.

3. E. M. Bender and G. Emerson, "Computational Linguistics and Grammar Engineering," in *Head-Driven Phrase Structure Grammar: The Handbook*, ed. S. Müller, A. Abeillé, R. D. Borsley, and J.-P. Koenig (Language Science Press, 2021), 1105–1153, https://doi.org/10.5281/zenodo.5599868.

4. N. Chomsky, *Aspects of the Theory of Syntax* (Cambridge, MA: MIT Press, 1965).

5. S. T. Piantadosi, "Modern Language Models Refute Chomsky's Approach to Language," *From Fieldwork to Linguistic Theory: A Tribute to Dan Everett* (Language Science Press, 2023), 353–414; L. S. Hamilton and A. G. Huth, "The Revolution Will Not Be Controlled: Natural Stimuli in Speech Neuroscience," *Language, Cognition and Neuroscience* 35, no. 5 (2018), https://doi.org/10.1080/23273798.2018.1499946.

6. A. Narayanan and S. Kapoor, "Evaluating LLMs Is a Minefield" (slide presentation, Princeton University, October 4, 2023), https://www.cs.princeton.edu/~arvindn/talks/evaluating_llms_minefield/.

7. N. Chomsky, *Knowledge of Language: Its Nature, Origin, and Use* (New York: Praeger Scientific, 1986).

8. W. von Humboldt, *On Language* (Cambridge: Cambridge University Press, 1999; originally published 1836).

9. G. Hinton, comment during panel discussion "Research on Intelligence in the Age of AI" Center for Brains, Minds, and Machines (October 7, 2023; clip posted on *Statistical Machine Learning* YouTube channel, https://www.youtube.com/watch?v=urBFz6-gHGY).

10. Piantadosi, "Modern Language Models Refute Chomsky's Approach to Language," 353–414.

11. M. Palmer, D. Gildea, and P. Kingsbury, "The Proposition Bank: An Annotated Corpus of Semantic Roles," *Computational Linguistics* 31, no. 1 (2005): 71–106.

12. M. Marcus, B. Santorini, and M. A. Marcinkiewicz, "Building a Large Annotated Corpus of English: The Penn Treebank," *Computational Linguistics* 19, no. 2 (1993): 313–330.

13. D. Jurafsky and J. H. Martin, *Speech and Language Processing* (third ed., forthcoming), chap. 22, https://web.stanford.edu/~jurafsky/slp3/; S. R. Ahmed, G. A. Baker, E. Judge, M. Regan, K. Wright-Bettner, M. Palmer, and J. H. Martin, "Linear Cross-Document Event Coreference Resolution with X-AMR" (preprint, March 25, 2024), https://doi.org/10.48550/arXiv.2404.08656.

14. G. Marcus, "Deep Learning: A Critical Appraisal" (preprint, January 2, 2018), https://doi.org/10.48550/arXiv.1801.00631.

15. G. Marcus, "Caricaturing Noam Chomsky" (blog post, March 11, 2023), https://garymarcus.substack.com/p/caricaturing-noam-chomsky.

16. Fromkin Speech Error Database, https://www.mpi.nl/dbmpi/sedb/sperco_form4.pl.

17. D. Fox and R. Katzir, "Large Language Models and Theoretical Linguistics," *Theoretical Linguistics* 50, nos. 1–2 (2024): 71–76.

18. V. H. Yngve, "A Model and an Hypothesis for Language Structure," *Proceedings of the American Philosophical Society* 104, no. 5 (1960): 444–466.

19. S. Piantadosi, "Modern Language Models Refute Chomsky's Approach to Language" (preprint, November 2023), https://ling.auf.net/lingbuzz/007180.

20. G. T. Fechner, *Elemente der Psychophysik* [Elements of Psychophysics] (Leipzig: Breitkopf und Härtel, 1860), vol. 2.

21. S. Dehaene, M. Piazza, P. Pinel, and L. Cohen, "Three Parietal Circuits for Number Processing," in *The Handbook of Mathematical Cognition*, ed. J. I. D. Campbell (Psychology Press, 2005), 433–453.

22. N. Chomsky, *Aspects of the Theory of Syntax* (MIT Press, 1965), 24–26.

23. D. Yamins and J. DiCarlo, "Using Goal-Driven Deep Learning Models to Understand Sensory Cortex," *Nature Neuroscience* 19 (2016): 356–365, https://doi.org/10.1038/nn.4244.

24. D. Pathak, P. Krähenbühl, and T. Darrell, "Constrained Convolutional Neural Networks for Weakly Supervised Segmentation," in *2015 IEEE International Conference on Computer Vision* (Institute of Electrical and Electronics Engineers, 2015), 1796–1804.

25. R. Cao and D. Yamins, "Explanatory Models in Neuroscience, Part 1: Taking Mechanistic Abstraction Seriously," *Cognitive Systems Research* 87 (2024): 101244.

26. A. Azulay and Y. Weiss, "Why Do Deep Convolutional Networks Generalize So Poorly to Small Image Transformations?" *Journal of Machine Learning Research* 20, no. 184 (2019): 1–25; O. Shifman and Y. Weiss, "Lost in Translation: Modern Neural Networks Still Struggle with Small Realistic Image Transformations" (preprint, April 10, 2024), https://doi.org/10.48550/arXiv.2404.07153.

27. "Hans Niemann Report," *Chess.com*, October 4, 2022, https://www.chess.com/blog/CHESScom/hans-niemann-report.

28. R. Lam, A. Sanchez-Gonzalez, M. Willson, P. Wirnsberger, M. Fortunato, F. Alet, S. Ravuri, T. Ewalds, Z. Eaton-Rosen, W. Hu, A. Merose, S. Hoyer, G. Holland, O. Vinyals, J. Stott, A. Pritzel, S. Mohamed, and P. Battaglia, "Learning Skillful Medium-Range Global Weather Forecasting," *Science* 382, no. 6677 (2023): 1416–1421.

29. Lam et al., "Learning Skillful Medium-Range Global Weather Forecasting."

Chapter 6

1. A. Krizhevsky, I. Sutskever, and G. E. Hinton, "ImageNet Classification with Deep Convolutional Neural Networks," *Advances in Neural Information Processing Systems 25 (NIPS 2012)*, ed. F. Pereira, C. J. Burges, L. Bottou, K. Q. Weinberger (Neural

Information Processing Systems Foundation, 2012), https://papers.nips.cc/paper_files/paper/2012.

2. I. J. Goodfellow, J. Shlens, and C. Szegedy, "Explaining and Harnessing Adversarial Examples" (preprint, December 20, 2014), https://doi.org/10.48550/arXiv.1412.6572.

3. A. Turing, "Computing Machinery and Intelligence," *Mind* 59, no. 236 (1950): 433–460.

4. H. J. Levesque, E. Davis, and L. Morgenstern, "The Winograd Schema Challenge," in *KR '12: Proceedings of the Thirteenth International Conference on Principles of Knowledge Representation and Reasoning*, ed. G. Brewka, T. Eiter, and S. A. McIlraith (Association for the Advancement of Artificial Intelligence, 2012), 552–561; H. J. Levesque, "On Our Best Behaviour," *Artificial Intelligence* 212 (2014): 27–35.

5. Levesque, "On Our Best Behaviour,"30.

6. T. Winograd, *Understanding Natural Language* (New York: Academic Press, 1972).

7. V. Kocijan, E. Davis, T. Lukasiewicz, G. Marcus, and L. Morgenstern, "The Defeat of the Winograd Schema Challenge," *Artificial Intelligence* (2023): 103971.

8. "Winograd Schema Challenge," *Wikipedia*, accessed June 23, 2024, https://en.wikipedia.org/wiki/Winograd_schema_challenge.

9. D. Büring, *Binding Theory* (Cambridge University Press, 2005).

10. D. Herel and T. Mikolov, "Advancing State of the Art in Language Modeling" (preprint, November 28, 2023), https://doi.org/10.48550/arXiv.2312.03735.

11. J. Kallini, I. Papadimitriou, R. Futrell, K. Mahowald, and C. Potts, "Mission: Impossible Language Models" (preprint, August 2, 2024), https://doi.org/10.48550/arXiv.2401.06416.

12. I. Sutskever (OpenAI chief scientist), "Why Next-Token Prediction Is Enough for AGI," interview excerpt, *Dwarkesh Podcast*, March 27, 2023, https://www.youtube.com/watch?v=YEUclZdj_Sc.

13. L. S. Hamilton and A. G. Huth, "The Revolution Will Not Be Controlled: Natural Stimuli in Speech Neuroscience," *Language, Cognition and Neuroscience* 35, no. 5 (2020): 573–582; A. Goldstein, Z. Zada, E. Buchnik, M. Schain, A. Price, B. Aubrey, S. A. Nastase, A. Feder, D. Emanuel, A. Cohen, A. Jansen, Harshvardhan Gazula, Gina Choe, Aditi Rao, Catherine Kim, Colton Casto, Lora Fanda, Werner Doyle, Daniel Friedman, Patricia Dugan, Lucia Melloni, Roi Reichart, Sasha Devore, Adeen Flinker, Liat Hasenfratz, Omer Levy, Avinatan Hassidim, Michael Brenner, Yossi Matias, Kenneth A. Norman, Orrin Devinsky, and Uri Hasson, "Shared Computational Principles for Language Processing in Humans and Deep Language Models," *Nature Neuroscience* 25, no. 3 (2022): 369–380.

14. N. Lan, E. Chemla, and R. Katzir, "Large Language Models and the Argument from the Poverty of the Stimulus," *Linguistic Inquiry* (2024): 1–28.

15. A. Wang, A. Singh, J. Michael, F. Hill, O. Levy, and S. R. Bowman, "GLUE: A Multi-Task Benchmark and Analysis Platform for Natural Language Understanding" (preprint, September 18, 2018), https://doi.org/10.48550/arXiv.1804.07461; A. Warstadt, A. Parrish, H. Liu, A. Mohananey, W. Peng, S. F. Wang, and S. R. Bowman, "BLiMP: The Benchmark of Linguistic Minimal Pairs for English," *Transactions of the Association for Computational Linguistics* 8 (2020): 377–392; A. Parrish, W. Huang, O. Agha, S. H. Lee, N. Nangia, A. Warstadt, K. Aggarwal, E. Allaway, T. Linzen, and S. R. Bowman, "Does Putting a Linguist in the Loop Improve NLU Data Collection?" (preprint, April 15, 2021), https://doi.org/10.48550/arXiv.2104.07179.

16. Warstadt et al., "BLiMP."

17. "AI and the Limits of Language," *Noēma*, August 23, 2022.

18. Y. Grodzinsky, I. Deschamps, and L. P. Shapiro, "Young Children and Patients with Broca's Aphasia Can Reconstruct Elided VPs," in *The Oxford Handbook of Ellipsis*, ed. J. van Craenenbroeck and T. Temmerman (Oxford University Press, 2019), 425–443.

19. A. Turing, "Computing Machinery and Intelligence," *Mind* 59, no. 236 (1950): 450.

20. "Morphophonemics of Modern Hebrew" (University of Pennsylvania, 1951).

Chapter 7

1. J. Cohen and B. West, *Gabby Giffords Won't Back Down* (2022), https://www.imdb.com/title/tt17490116/.

2. P. Broca, "Remarques sur le siège de la faculté du langage articulé, suivies d'une observation d'aphémie (perte de la parole)" [Remarks on the Seat of the Faculty of Articulated Language, Following an Observation of Aphemia (Loss of Speech)], *Bulletin de la Société Anatomique* 6, nos. 330–357 (1861): 27. English translation in *Broca's Region*, ed. Y. Grodzinsky and K. Amunts.

3. N. Geschwind, "The Organization of Language and the Brain: Language Disorders After Brain Damage Help in Elucidating the Neural Basis of Verbal Behavior," *Science* 170, no. 3961 (1970): 940–944.

4. L. Lichtheim, "On Aphasia," *Brain* (1885). Reprinted in *Broca's Region*, ed. Y. Grodzinsky and K Amunts ((New York: Oxford University Press, 2006), 318–347.

5. S. J. Gould, *Mismeasure of Man* (Norton, 1981).

6. R. Jakobson, *Kindersprache, Aphasie und allgemeine Lautgesetze* (Uppsala: Almqvist & Wiksell, 1941). Translated into English as *Child Language, Aphasia, and Phonological Universals* (The Hague: Mouton, 1968).

7. S. A. Blumstein, *A Phonological Investigation of Aphasic Speech* (Walter de Gruyter, 2018).

8. N. Mesgarani, C. Cheung, K. Johnson, and E. F. Chang, "Phonetic Feature Encoding in Human Superior Temporal Gyrus," *Science* 343, no. 6174 (2014): 1006–1010.

9. C. Tang, L. S. Hamilton, and E. F. Chang, "Intonational Speech Prosody Encoding in the Human Auditory Cortex," *Science* 357, no. 6353 (2017): 797–801, https://doi.org/10.1126/science.aam8577; A. Mai, S. Riès, S. Ben-Haim, J. J. Shih, and T. Q. Gentner, "Acoustic and Language-Specific Sources for Phonemic Abstraction from Speech," *Nature Communications* 15 (2024): 677, https://doi.org/10.1038/s41467-024-44844-9.

10. Y. Grodzinsky and I. Nelken, "The Neural Code That Makes Us Human". Science, 343 (2014): 978-9, DOI: 10.1126/science.1251495.

11. Y. Grodzinsky, "Language Deficits and the Theory of Syntax," *Brain and Language* 27, no. 1 (1986): 135–159; "The Neurology of Syntax: Language Use Without Broca's Area," *Behavioral and Brain Sciences* 23, no. 1 (2000): 1–21.

12. G. Hickok and S. Avrutin, "Comprehension of Wh-Questions in Two Broca's Aphasics," *Brain and Language* 52, no. 2 (1996): 314–327.

13. K. Stromswold, D. Caplan, N. Alpert, and S. Rauch, "Localization of Syntactic Comprehension by Positron Emission Tomography," *Brain and Language* 52, no. 3 (1996): 452–473; M. Ben-Shachar, D. Palti, and Y. Grodzinsky, "Neural Correlates of Syntactic Movement: Converging Evidence from Two fMRI Experiments," *NeuroImage* 21, no. 4 (2004): 1320–1336; A. D. Friederici, C. J. Fiebach, M. Schlesewsky, I. D. Bornkessel, and D. Y. von Cramon, "Processing Linguistic Complexity and Grammaticality in the Left Frontal Cortex," *Cerebral Cortex* 16, no. 12 (2006): 1709–1717.

14. A. Santi and Y. Grodzinsky, "Working Memory and Syntax Interact in Broca's Area," *NeuroImage* 37, no. 1 (2007): 8–17.

15. A. Santi and Y. Grodzinsky, "fMRI Adaptation Dissociates Syntactic Complexity Dimensions," *NeuroImage* 51, no. 4 (2010), 1285–1293.

16. M. Makuuchi, Y. Grodzinsky, K. Amunts, A. Santi, and A. Friederici, "Processing Noncanonical Sentences in Broca's Region: Reflections of Movement Distance and Type," *Cerebral Cortex* 23, no. 3 (2013): 694–702; Y. Grodzinsky and A. Santi, "The Battle for Broca's Region," *Trends in Cognitive Sciences* 12, no. 12 (2008): 474–480.

17. M. Riva, S. M. Wilson, R. Cai, A. Castellano, K. M. Jordan, R. G. Henry, M. L. G. Tempini, M. S. Berger, and E. F. Chang, "Evaluating Syntactic Comprehension

During Awake Intraoperative Cortical Stimulation Mapping," *Journal of Neurosurgery* 138, no. 5 (2022): 1403–1410.

18. A. Moro, *The Boundaries of Babel: The Brain and the Enigma of Impossible Languages* (MIT Press, 2015); A. D. Friederici, *Language in Our Brain: The Origins of a Uniquely Human Capacity* (MIT Press, 2017); G. Baggio, *Meaning in the Brain* (MIT Press, 2018).

19. Y. Grodzinsky, I. Deschamps, P. Pieperhoff, F. Iannilli, G. Agmon, Y. Loewenstein, and K. Amunts, "Logical Negation Mapped onto the Brain," *Brain Structure and Function* 225 (2020): 19–31.

20. "The Nobel Prize in Physiology or Medicine 1906," *NobelPrize.org* (n.d.), https://www.nobelprize.org/prizes/medicine/1906/summary/.

21. B. Fischl and M. I. Sereno, "Microstructural Parcellation of the Human Brain," *NeuroImage* 182 (2018): 219–231.

22. K. Zilles and K. Amunts, "Centenary of Brodmann's Map—Conception and Fate," *Nature Reviews Neuroscience* 11, no. 2 (2010): 139–145.

23. K. Amunts, A. Schleicher, U. Bürgel, H. Mohlberg, H. B. Uylings, and K. Zilles, "Broca's Region Revisited: Cytoarchitecture and Intersubject Variability," *Journal of Comparative Neurology* 412, no. 2 (1999): 319–341.

24. K. Amunts, C. Lepage, L. Borgeat, H. Mohlberg, T. Dickscheid, M. É. Rousseau, S. Bludau, P. L. Bazin, L. B. Lewis, A. M. Oros-Peusquens, and N. J. Shah, "BigBrain: An Ultrahigh-Resolution 3D Human Brain Model," *Science* 340, no. 6139 (2013): 1472–1475.

25. Y. Grodzinsky, I. Deschamps, P. Pieperhoff, F. Iannilli, G. Agmon, Y. Loewenstein, and K. Amunts, "Logical Negation Mapped onto the Brain," *Brain Structure and Function* 225 (2020): 19–31.

26. M. M. Monti, D. N. Osherson, M. J. Martinez, and L. M. Parsons, "Functional Neuroanatomy of Deductive Inference: A Language-Independent Distributed Network," *NeuroImage* 37 (2007): 1005–1016.

27. W. Matchin and G. Hickok, "The Cortical Organization of Syntax," *Cerebral Cortex* 30, no. 3 (2020): 1481–1498.

28. Matchin and Hickok, "The Cortical Organization of Syntax."

29. Y. Grodzinsky, P. Pieperhoff, and C. Thompson, "Stable Brain Loci for the Processing of Complex Syntax: A Review of the Current Neuroimaging Evidence," *Cortex* 142 (2021): 252–271.

30. N. Saadon-Grosman, Y. Loewenstein, and S. Arzy, "The 'Creatures' of the Human Cortical Somatosensory System," *Brain Communications* 2, no. 1 (2020): fcaa003.

31. A. Kohn, A. I. Jasper, J. D. Semedo, E. Gokcen, C. K. Machens, and M. Y. Byron, "Principles of Corticocortical Communication: Proposed Schemes and Design Considerations," *Trends in Neurosciences* 43, no. 9 (2020): 725–737.

32. (MIT Press, 1983).

Chapter 8

1. T. Serre, A. Oliva, and T. Poggio, "A Feedforward Architecture Accounts for Rapid Categorization," *Proceedings of the National Academy of Sciences of the United States of America* 104 (2007): 6424–6429.

2. D. L. Yamins and J. J. DiCarlo, "Using Goal-Driven Deep Learning Models to Understand Sensory Cortex," *Nature Neuroscience* 19, no. 3 (2016): 356–365.

3. E. Margalit, H. Lee, D. Finzi, J. J. DiCarlo, K. Grill-Spector, and D. L. Yamins, "A Unifying Framework for Functional Organization in Early and Higher Ventral Visual Cortex," *Neuron* 112, no. 14 (2024): 2435–2451.

4. A. Doerig, R. P. Sommers, K. Seeliger, B. Richards, J. Ismael, G. W. Lindsay, K. P. Kording, T. Konkle, M. A. J. van Gerven, N. Kriegeskorte, and T. C. Kietzmann, "The Neuroconnectionist Research Programme," *Nature Reviews Neuroscience* 24, no. 7 (2023): 431–450; R. Cao and D. Yamins, "Explanatory Models in Neuroscience, Part 1: Taking Mechanistic Abstraction Seriously," *Cognitive Systems Research* 87 (2024): 101244.

5. A. Goldstein, Z. Zada, E. Buchnik, M. Schain, A. Price, B. Aubrey, S. A. Nastase, A. Feder, D. Emanuel, A. Cohen, A. Jansen, H. Gazula, G. Choe, A. Rao, C. Kim, C. Casto, L. Fanda, W. Doyle, D. Friedman, P. Dugan, L. Melloni, R. Reichart, S. Devore, A. Flinker, L. Hasenfratz, O. Levy, A. Hassidim, M. Brenner, Y. Matias, K. A. Norman, O. Devinsky, and U. Hasson, "Shared Computational Principles for Language Processing in Humans and Deep Language Models," *Nature Neuroscience* 25, no. 3 (2022): 369–380.

6. A. Goldstein, A. Grinstein-Dabush, M. Schain, H. Wang, Z. Hong, B. Aubrey, S. A. Nastase, Z. Zada, E. Ham, A. Feder, H. Gazula, E. Buchnik, W. Doyle, S. Devore, P. Dugan, R. Reichart, D. Friedman, M. Brenner, A. Hassidim, O. Devinsky, A. Flinker, and U. Hasson, "Alignment of Brain Embeddings and Artificial Contextual Embeddings in Natural Language Points to Common Geometric Patterns," *Nature Communications* 15, no. 1 (2024): 2768.

7. M. Schrimpf, I. A. Blank, G. Tuckute, C. Kauf, E. A. Hosseini, N. Kanwisher, J. B. Tenenbaum, and E. Fedorenko, "The Neural Architecture of Language: Integrative Modeling Converges on Predictive Processing," *Proceedings of the National Academy of Sciences* 118, no. 45 (2021): e2105646118; K. Mahowald, A. A. Ivanova, I. A. Blank, N. Kanwisher, J. B. Tenenbaum, and E. Fedorenko, "Dissociating Language and Thought in Large Language Models," *Trends in Cognitive Sciences* 28, no. 6 (2024): 517–540.

8. N. Kanwisher, "Domain Specificity in Face Perception," *Nature Neuroscience* 3, no. 8 (2000): 759–763.

9. M. Schrimpf, I. A. Blank, G. Tuckute, C. Kauf, E. A. Hosseini, N. Kanwisher, J. B. Tenenbaum, and E. Fedorenko, "The Neural Architecture of Language: Integrative Modeling Converges on Predictive Processing," *Proceedings of the National Academy of Sciences* 118, no. 45 (2021): e2105646118.

10. Y. Grodzinsky, "The Picture of the Linguistic Brain: How Sharp Can It Be? Reply to Fedorenko & Kanwisher," *Language and Linguistics Compass* 4, no. 8 (2010): 605–622; E. Fedorenko, P. J. Hsieh, A. Nieto-Castañón, S. Whitfield-Gabrieli, and N. Kanwisher, "New Method for fMRI Investigations of Language: Defining ROIs Functionally in Individual Subjects," *Journal of Neurophysiology* 104, no. 2 (2010), 1177–1194.

11. K. Mahowald, A. A. Ivanova, I. A. Blank, N. Kanwisher, J. B. Tenenbaum, and E. Fedorenko, "Dissociating Language and Thought in Large Language Models," *Trends in Cognitive Sciences* 28, no. 6 (2024): 517–540; G. Tuckute, A. Sathe, S. Srikant, M. Taliaferro, M. Wang, M. Schrimpf, K. Kay, and E. Fedorenko, "Driving and Suppressing the Human Language Network Using Large Language Models," *Nature Human Behaviour* 8, no. 3 (2024): 544–561.

12. D. Fox and R. Katzir, "Large Language Models and Theoretical Linguistics," *Theoretical Linguistics* 50, nos. 1–2 (2024): 71–76.

13. A. Warstadt, A. Parrish, H. Liu, A. Mohananey, W. Peng, S. F. Wang, and S. R. Bowman, "BLiMP: The Benchmark of Linguistic Minimal Pairs for English," *Transactions of the Association for Computational Linguistics* 8 (2020): 377–392.

Epilogue

1. J. Kathryn Bock, "Syntactic Persistence in Language Production," *Cognitive Psychology* 18, no. 3 (1986): 355–387; Kristen M. Tooley and Matthew J. Traxler, "Syntactic Priming Effects in Comprehension: A Critical Review," *Language and Linguistics Compass* 4, no. 10 (2010): 925–937.

2. F. Chang, G. S. Dell, and K. Bock, "Becoming Syntactic," *Psychological Review* 113, no. 2 (2006): 234–272, https://doi.org/10.1037/0033-295X.113.2.234; A. Santi and Y. Grodzinsky, "fMRI Adaptation Dissociates Syntactic Complexity Dimensions," *NeuroImage* 51, no. 4 (2010): 1285–1293.

3. T. J. Sejnowski, "Large Language Models and the Reverse Turing Test," *Neural Computation* 35, no. 3 (2023): 309–342.

4. Brigadier General Y. S., *The Human Machine Team* (eBookPro Publishing, 2021), 104; Lieutenant Colonel (res.) G. and Major (res.) L., "On SIGINT and AI," Dado Center, Israel Defense Forces, January 24, 2021.

5. N. Dvori, Mako TV (Israel), May 19, 2021.

6. U. Bar-Joseph, *The Watchman Fell Asleep: The Surprise of Yom Kippur and Its Sources* (SUNY Press, 2005).

7. U. Bar-Joseph, "The Complacency Leading to the War Stemmed Not Only from an Intelligence Failure," *Ha'aretz*, November 1, 2023.

8. Y. Azizi, "The Things You See from There," *Ha'aretz*, November 23, 2023; Channel 12 (Mako) News (Israel), December 3, 2023.

9. B. Caspit, "8200's Concealed Report About the Failures Leading to the Massacre in the South," *Ma'ariv*, February 24, 2024.

Index

Publisher contact:
The MIT Press
Massachusetts Institute of Technology
77 Massachusetts Avenue, Cambridge, MA 02139
mitpress.mit.edu

EU Authorised Representative:
Easy Access System Europe, Mustamäe tee 50,
10621 Tallinn, Estonia
gpsr.requests@easproject.com

Printed by Integrated Books International,
United States of America